LOW-POWER DEEP SUB-MICRON CMOS LOGIC

THE KLUWER INTERNATIONAL SERIES IN ENGINEERING AND COMPUTER SCIENCE

ANALOG CIRCUITS AND SIGNAL PROCESSING
Consulting Editor: Mohammed Ismail. *Ohio State University*

LOW-POWER DEEP SUB-MICRON CMOS LOGIC

Sub-threshold Current Reduction

By

P.R. van der Meer

Delft University of Technology,
Delft, The Netherlands

and

A. van Staveren

National Semiconductor Corporation,
Delft, The Netherlands

and

A.H.M. van Roermund

Eindhoven University of Technology,
Eindhoven, The Netherlands

KLUWER ACADEMIC PUBLISHERS

BOSTON / DORDRECHT / LONDON

A C.I.P. Catalogue record for this book is available from the Library of Congress.

ISBN 978-1-4757-1057-1   ISBN 978-1-4020-2849-6 (eBook)
DOI 10.1007/978-1-4020-2849-6

Published by Kluwer Academic Publishers,
P.O. Box 17, 3300 AA Dordrecht, The Netherlands.

Sold and distributed in North, Central and South America
by Kluwer Academic Publishers,
101 Philip Drive, Norwell, MA 02061, U.S.A.

In all other countries, sold and distributed
by Kluwer Academic Publishers,
P.O. Box 322, 3300 AH Dordrecht, The Netherlands.

Printed on acid-free paper

Contents

Index of symbols

Symbol	Meaning	Unit
α	Node transition-cycle activity factor	
α_{ij}	Node transition-cycle activity factor α of node j in clock period i	
β	Effective transistor strength	$A \cdot V^{-2}$
γ	Body factor	$V^{-\frac{1}{2}}$
δ	Relative increase of the logic gate capacitance	
ϵ	Unit step function	
η	Reversibility factor	
θ	Phase angle between voltage and current	radians
θ_t	STI transition angle	$^\circ$
ϑ	DIBL factor	
κ	Inverse scaling factor	
μ_0	Zero bias mobility	$m^2 \cdot V^{-1}s^{-1}$
μ_n	Electron mobility	$m^2 \cdot V^{-1}s^{-1}$
μ_p	Hole mobility	$m^2 \cdot V^{-1}s^{-1}$
ν	Carrier mobility temperature exponent	
σ	Drain-source saturation current exponent	
τ	Characteristic time; product of R and C	s
τ_d	Time delay	s
τ_f	Fall time	s
τ_l	Carrier lifetime	s
τ_{pd}	Propagation delay	s
τ_r	Rise time	s
ϕ_F	Fermi potential	V
ϕ_{gc}	Gate-channel work-function difference	V
χ	Threshold voltage temperature coefficient	$V \cdot K^{-1}$

ψ_s	Surface potential	V
ω_0	Oscillation frequency	s^{-1}
đ	Diminutive	
e_{C_T}	Instantaneous energy of the tank capacitance	J
f_{clk}	Clock frequency	s^{-1}
h	Isolation height with respect to the silicon surface	m
i_{C_L}	Instantaneous current through the load (node) capacitance	A
i_{sc}	Instantaneous short-circuit current	A
k	Boltzmann's constant, $1.380658 \cdot 10^{-23}$	$J \cdot K^{-1}$
n	Slope factor	
n_a	Number of atoms	
n_i	Intrinsic concentration of electrons	m^{-3}
n_{p0}	Thermal-equilibrium concentration of minority electrons in a p-type region	m^{-3}
p	Number of voltage steps in the step-wise charging technique	
$p_{dn,ij}$	Instantaneous power delivered to the power supply to discharge node j in clock period i	W
p_{ij}	Instantaneous power delivered by or to the power supply to charge or discharge node j in clock period i	W
p_{n0}	Thermal-equilibrium concentration of minority holes in an n-type region	m^{-3}
$p_{up,ij}$	Instantaneous power delivered by the power supply to charge node j in clock period i	W
q	Electron charge, $1.602177 \cdot 10^{-19}$	C
r	STI corner radius	m
s_w	Wire spacing	m
t_{dep}	Depletion region thickness	m
t_i	Clock period i	s
t_j	Junction depth	m
t_{ox}	Gate oxide thickness	m
t_w	Wire thickness	m
u_{C_L}	Instantaneous voltage across the load (node) capacitance	V
u_r	Instantaneous voltage across a resistor	V
u_{sw}	Instantaneous logic voltage swing	V
A_g	Gate area	m^2
A_j	Source-bulk or drain-bulk pn-junction area	m^2
A_w	Wiring area	m^2

C	Capacitance	F
C_{clk_std}	Switched capacitance of the clock circuitry of a standard circuit	F
C_{clk_TS}	Switched capacitance of the clock circuitry of a Triple-S circuit	F
C'_d	Depletion capacitance per unit area	$F \cdot m^{-2}$
C_{data_std}	Switched capacitance of a standard circuit, when all data bits change state every clock cycle	F
C_{data_TS}	Switched capacitance of a Triple-S circuit, when all data bits change state every clock cycle	F
C_g	Gate capacitance	F
C_j	Junction capacitance	F
C_L	Load (node) capacitance	F
C_{node}	Circuit node capacitance	F
C_{ox}	Gate oxide capacitance	F
C'_{ox}	Gate oxide capacitance per unit area	$F \cdot m^{-2}$
C_{pad}	Bond-pad capacitance	F
C_s	Series capacitance	F
C_T	Tank capacitance	F
C'_{th}	Heat capacitance per unit area	$J \cdot K^{-1} \cdot m^{-2}$
C_w	Wiring capacitance	F
D_n	Electron diffusion coefficient	$m^2 \cdot s^{-1}$
D_p	Hole diffusion coefficient	$m^2 \cdot s^{-1}$
$E_{battery}$	Energy stored in a battery	$A \cdot h \; or \; J$
E_C	Conduction band energy (top edge)	$J \; or \; eV$
E_{C_L}	Energy stored in the load capacitor	J
E_{cycle}	Transition-cycle energy	J
E_{dn}	Energy delivered to the power supply to discharge a circuit node	J
$E_{dn,ij}$	Energy delivered to the power supply to discharge node j in clock period i	J
E_F	Fermi energy	$J \; or \; eV$
E_{Fi}	Intrinsic Fermi energy	$J \; or \; eV$
E_{func}	Functional energy	J
E_g	Band gap energy	$J \; or \; eV$
E_i	Initial energy	J
E_{ins}	Electric field in the gate insulator	$V \cdot m^{-1}$
E_{int}	Internal energy of a process	J
$E_{overhead}$	Excess energy consumption	J
E_{ox}	Electric field in the gate oxide	$V \cdot m^{-1}$

E_R	Energy dissipated in a resistor	J
E_{saved}	Energy saved by switching into standby	J
E_{up}	Energy delivered by the power supply to charge a circuit node	J
$E_{up,ij}$	Energy delivered by the power supply to charge node j in clock period i	J
E_V	Valence band energy (bottom edge)	$J\ or\ eV$
G'_{conv}	Thermal convection	$W \cdot K^{-1} \cdot m^{-2}$
GND	Ground	V
G'_{th}	Thermal conductance	$W \cdot K^{-1} \cdot m^{-2}$
I_0^*	Leakage current per unit width	$A \cdot m^{-1}$
I_1	Channel leakage current	A
I_2	Diode leakage current	A
I_3	Gate leakage current	A
I_d	Drain current	A
I_{ds}	Drain-source current	A
$I_{ds,sat}$	Drain-source saturation current	A
I_{gc}	Gate-channel tunneling current	A
I_{gd}	Drain extension to gate overlap tunneling current current	A
I_{GIDL}	Gate induced drain leakage current	A
I_{gs}	Source extension to gate overlap tunneling current	A
$I_{leak,max}$	Maximum leakage current of a circuit	A
$I_{wi,U_{th,low}}$	Weak-inversion current of low-threshold-voltage transistors	A
J_{FN}	Fowler-Nordheim tunneling current	A
J_{gen}	Generation current density	$A \cdot m^{-2}$
J_s	Reverse saturation current density	$A \cdot m^{-2}$
K	Permittivity	$F \cdot m^{-1}$
K_0	Permittivity of free space, $8.85 \cdot 10^{-12}$	$F \cdot m^{-1}$
K_{ID}	Permittivity of the isolation dielectric material	$F \cdot m^{-1}$
K_{ins}	Permittivity of the gate insulator material	$F \cdot m^{-1}$
K_{ox}	Permittivity of gate oxide (SiO_2)	$F \cdot m^{-1}$
K_{Si}	Permittivity of Silicon	$F \cdot m^{-1}$
L	Inductance	H
L_{ch}	MOS channel length	m
L_n	Minority carrier electron diffusion length	m
L_p	Minority carrier hole diffusion length	m
L_s	Source junction length	m
L_w	Wire length	m

N	Number of circuit nodes	
N_a	Acceptor doping concentration	m^{-3}
N_d	Donor doping concentration	m^{-3}
N_{ss}	Oxide fixed charge density	m^{-3}
N_{sub}	Substrate doping concentration	m^{-3}
P	Pressure	$N \cdot m^{-2}$
$\overline{P}_{active}$	Average power over all active modes	W
P_{el}	Electrical power	W
P_{ext}	External pressure	$N \cdot m^{-2}$
P_f	Final external pressure	$N \cdot m^{-2}$
P_{func}	Functional power	W
$P_{func\_std,0}$	Functional power of a standard circuit under zero data activity conditions	W
$P_{func\_std,max}$	Functional power of a standard circuit under maximum data activity conditions	W
$P_{func\_TS,max}$	Functional power of a Triple-S circuit under maximum data activity conditions	W
P'_{heat}	Heat flow	$W \cdot m^{-2}$
P_i	Initial external pressure	$N \cdot m^{-2}$
$P_{short-circuit}$	Parasitical power	W
$\overline{P}_{standby}$	Average power over all standby modes	W
Q	Charge	C
Q'_d	Charge density induced in the drain electrode	$C \cdot m^{-2}$
Q_f	Quality factor	
Q_{gate}	Gate charge	C
Q_h	Heat	J
Q_{ss}	Oxide fixed charge	C
R	Resistance	Ω
R_{ds}	MOS channel impedance	Ω
Red	Leakage reduction factor	
$R^*_{psw,on}$	Power switch on-impedance times unit width	$\Omega \cdot m$
S	Entropy	$J \cdot K^{-1}$
S_{wi}	Weak-inversion "slope"	$mV \cdot decade^{-1}$
T	Temperature	K
T_{active}	Active period time	s
$T_{active,max}$	Max. total active mode time	s
T_{amb}	Ambient temperature	K
T_{clk}	Clock period time	s
$T_{lifetime}$	Battery lifetime	s
T_r	Room temperature	K
$T_{standby}$	Standby period time	s
$T_{standby,max}$	Max. total standby mode time	s

T_{stdb,be_oh}	Break-even standby time in case of overhead costs	s
T_{stdb,be_se}	Break-even standby time in case of stored energy	s
U	Voltage	V
U_b	Bulk (substrate) voltage	V
U_{bi}	Built-in potential barrier	V
U_d	Drain voltage	V
U_{db}	Drain-bulk voltage	V
U_{dd}	Supply voltage	V
U_{ddv}	Virtual supply voltage	V
U_{ds}	Drain-source voltage	V
U_{fb}	Flat-band voltage	V
$U_{fb,gc}$	Channel flat-band voltage	V
$U_{fb,sde}$	Source-drain extension to gate flat-band voltage	V
U_g	Gate voltage	V
U_{gb}	Gate-bulk voltage	V
U_{gs}	Gate-source voltage	V
U_H	Boundary potential between moderate inversion and strong inversion	V
U_L	Boundary potential between depletion and weak inversion	V
U_M	Boundary potential between weak inversion and moderate inversion	V
U_{psw}	Voltage across the power switch	V
U_s	Source voltage	V
U_{sb}	Source-bulk voltage	V
U_{sw}	Logic voltage swing	V
U_T	Boltzmann voltage	V
U_{th}	Threshold voltage	V
U_{th}^*	Effective threshold voltage	V
U_{th0}	Zero-bias threshold voltage	V
$U_{th,psw}$	Power switch threshold voltage	V
V	Volume	m^3
W	Work	J
W_{ch}	MOS channel width	m
W_{psw}	Width of the power switch	m
W_w	Wire width	m

1

INTRODUCTION

1.1 Power-dissipation trends in CMOS circuits

Shrinking device geometry, growing chip area and increased data-processing speed performance are technological trends in the integrated circuit industry to enlarge chip functionality. Already in 1965 Gordon Moore predicted that the total number of devices on a chip would double every year until the 1970s and every 24 months in the 1980s. This prediction is widely known as "Moore's Law" and eventually culminated in the Semiconductor Industry Association (SIA) technology road map [1]. The SIA road map has been a guide for the industry leading them to continued wafer and die size growth, increased transistor density and operating frequencies, and defect density reduction. To mention a few numbers; the die size increased 7% per year, the smallest feature sizes decreased 30% and the operating frequencies doubled every two years. As a consequence of these trends both the number of transistors and the power dissipation per unit area increase. In the near future the maximum power dissipation per unit area will be reached.

Down-scaling of the supply voltage is not only the most effective way to reduce power dissipation in general it also is a necessary precondition to ensure device reliability by reducing electrical fields and device temperature, to prevent device degradation. A draw-back of this solution is an increased signal propagation delay, which results in a lower data-processing speed performance. To get around this obstacle in complementary metal oxide semiconductor (CMOS) technology a possibility is to decrease the transistor's threshold voltage. However, lower threshold voltages result in increased channel-leakage currents in switched-off devices.

The number of battery-operated portable and wireless applications grows rapidly. Leakage power dissipation is already a serious problem in present

1

and increases for future portable systems, especially for those possessing a relatively long standby period compared to their active period. Consequently future portable digital systems are facing an increasing conflict between speed performance and channel-leakage power dissipation.

1.2   Overview of present power-reduction solutions

The power dissipation of a system can be influenced at several design levels like the system, algorithm, data, circuit, device and the technology level. To explore the field of power reduction the following literature overview will present a non-exhaustive list of power-reduction techniques for each design level.

The choice of algorithm determines power dissipation at the highest level of decision. At this level minimizing the number of operations necessary to perform a given function reduces the overall switching activity and therefore the power dissipation. A vivid example is for instance the way of finding a data element in a fixed list of data elements. In stead of accessing all entries (n) in the list from the first to the last during a linear search, a binary search can be performed when all the elements of the list have been arranged by one of their search attributes, e.g. magnitude, which reduces the number of accesses from $O(n)$ to $O(^2log(n))$ [2, 3, 4].

The correlation between successive data elements can be exploited through coding to reduce the switching activity at the system level. For instance in a Gray code sequence the successive elements vary only one bit and therefore are highly correlated. On the one hand application of Gray coding for addressing instructions in program memory reduces the number of transitions significantly, because successive address words are highly correlated [5]. On the other hand resource sharing, e.g. in time-multiplexed address and data busses or time-multiplexed execution units, disturbs the correlation between successive elements and therefore increases the switching activity again.

Since voltage scaling is the most effective approach to reduce power dissipation, different parts of a system can be operated on different fixed supply voltages depending on the speed requirements [6, 7].

Hardware parallelisation and pipelining provide for the necessary compensation in speed performance while the supply voltage is scaled statically [8]. Area is traded for lower power dissipation. The counter part of static voltage scaling is dynamic voltage control, which means that the supply voltage is controlled as a function of parameter variations [9] and workload [10]. In the case of a resonant circuit, for adiabatic switching [11], the supply voltage varies periodically. Parameter variations, which influence the performance, can be compensated for, whereas for workload changes the performance can be adjusted accordingly by adapting the supply voltage dynamically. When the power supply source and the circuit form a resonating system, this source delivers energy to the circuit to perform a basic operation, i.e. change the state of the circuit, and afterwards the stored energy is returned again to the power supply. In this way adiabatic switching is realized.

While the supply-voltage is reduced in favor of reducing the power dissipation still the speed performance has to be increased for each new generation to improve functionality of the application. At the device and technology level techniques have been developed to accomplish this goal. At the device level substrate biasing is applied to adjust the threshold voltage level of an MOS transistor depending on the mode of operation [12, 13]. In the active mode the speed performance has to be high and the thresholds therefore low, whereas in the standby mode the channel leakage has to be low and the thresholds high. At the device level the intrinsic threshold voltage level can be adapted by adjusting the channel doping. In a multi-threshold voltage process high-threshold voltage transistors are placed in non-critical paths, which will make them (almost) critical and low power, whereas low-threshold voltage transistors are placed in critical paths, which enhances the performance [14].

1.3 Aim and scope of this book

This book aims for a systematic approach to the reduction of power dissipation in deep sub-micron CMOS logic. From the global literature overview of the preceding paragraph it can be concluded that already a lot of effort has been put into finding power reduction techniques at the different levels. However, present literature lacks a systematic classification specifying all the sources of power dissipation in CMOS and the techniques to reduce these sources. A clear classification simplifies tracing blank spots, which may lead to new combinations or even new solutions. In this book the reduction of weak-inversion power dissipation will be emphasized, because ever increasing weak-inversion currents becomes a real threat to portable future applications using MOS technologies. Moreover this field has hardly been explored. Therefore, special attention will be paid to the classification of this type of power dissipation and reduction techniques.

The scope of this book covers a classification of the sources of power dissipation in deep sub-micron CMOS technologies. For each power-dissipation source the key parameters will be presented as well as techniques which influence them in order to reduce that specific source. In this classification a distinction has been made between power dissipated in the favor of information processing and power that is parasitical to that. The weak-inversion power dissipation is a subset of the latter category. Techniques reducing this parasitical power component as well as their effectiveness will come into focus. In the final part of this book a new weak-inversion current reduction technique is treated. In addition, measurement results will be presented and the applicability and effectiveness of this new technique in real applications is considered.

1.4   Organization of the book

In chapter 2 power and energy will be discussed as figures of merit for digital systems. Energy is a good figure of merit regarding power-reduction techniques with respect to battery lifetime. Neglecting time aspects means neglecting system dynamics, which could introduce errors in the power-dissipation and reduction observations. Power is more suitable in the case of determining the maximum chip temperature. Chapter 3 discusses a classification of power dissipation sources in digital CMOS circuits. The functional and parasitical power dissipation are the two main categories that have been distinguished. Trends in both classes of power dissipation will be presented and indicate the necessity for power reduction techniques. Chapter 4 and 5 consider power reduction methods for both the functional and the parasitical power dissipation categories, respectively. Chapter 5 will stress the need for weak-inversion current-reduction techniques especially for battery operated systems. Chapter 6 presents a classification of weak-inversion current-reduction techniques, which will be used to introduce and device a new weak-inversion current-reduction technique, called Triple-S. Chapter 7 considers the effectiveness of weak-inversion current-reduction techniques in general and for both existing and the Triple-S techniques. Chapter 8 considers necessary adaptations of the standard CMOS process flow to implement the Triple-S leakage-power reduction technique. Experimental designs containing Triple-S will be used to determine speed performance, i.e. propagation delays, occupation of chip area, functional power dissipation and leakage reduction and compare these items with standard designs, i.e. designs without leakage power reduction circuitry. Chapter 9, the last chapter of this book, presents the overall conclusions.

2

POWER VERSUS ENERGY

Although power consumption always has been an important issue considering electronic equipment design, nowadays it literally is a hot item. To address this issue in a complete way, power as well as energy consumption should be taken into account. Although power and energy are related via a straightforward time integral, the impact on a system when constraining it via power or energy consumption, can be completely different. Power is regarded when considering for instance heat production, on-chip and off-chip power supply circuits and the environment. In the case of heat production it may be clear that local power dissipation (power density) is a key parameter. Tremendously increasing power densities push chip temperatures to the limit. Energy is a key parameter when considering battery lifetime, energy supply by power plants and energy costs.

Section 2.1 considers the causes of the increased power consumption for each new technology generation and the consequences. Since batteries contain a limited amount of energy, battery lifetime is a main issue for mobile applications. Section 2.2 discusses the link between key parameters of mobile equipment and energy consumption.

2.1   Power considerations

The world wide market for micro electronics has an almost insatiable hunger for increased functionality per unit area and time. Reduction of feature sizes and increased clock frequencies are a few means to appease this hunger by increasing the switching activity per unit area and time. Although smaller feature sizes reduce the power consumption of individual node transitions, the increased power densities caused by increased switching frequencies and tran-

5

sistor densities surpass these reductions amply. Therefore, both the total power consumption of a chip and the power density will increase.

The increase of the total power consumption per chip has many consequences both on-chip and off-chip. The transport of power onto a complex data processing chip dissipating about 200 Watts at a supply voltage level of 1 Volt is a technological challenge. The number of pins devoted to just transporting energy will have to grow significantly, adding to the costs of packaging and printed circuit boards, while introducing a shorter mean time between failures. Power supplies will have to cope with ever increasing power demands. The demand for more power not only increases system and operational costs, it also burdens the environment, especially for battery operated systems.

The introduction of a standby mode to reduce the power consumption of a system in idle periods is already commonplace in portable systems and is also applied nowadays in desktop applications. Although the latter can drain the power needed from the wall outlet, standby modes can provide for a significant power and cost reduction, thereby relieving the environment as well. Figure 2.1 shows an example of active and standby cycles of desktop and laptop applications.

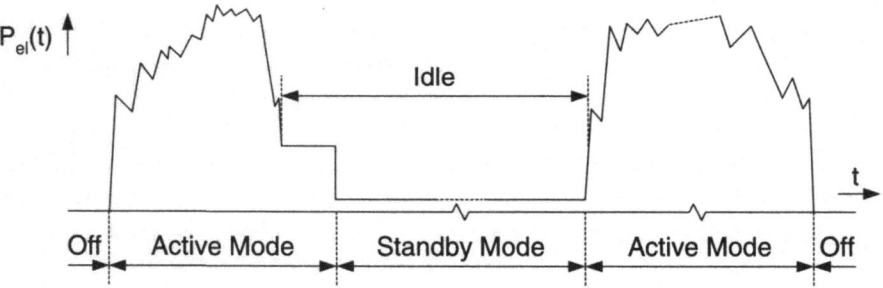

Figure 2.1.   Operation modes of desktop and laptop applications.

The increase of a chip's power density has serious consequences for the on-chip temperature management. Figure 2.2 shows the network equivalent of the first order thermal model of a chip and its environment. The source $P'_{heat}(Wm^{-2})$ [1] represents the heat flow originating from the power dissipation produced by the internal circuits. The heat conduction from the chip via the package to the ambient is modeled by the thermal conductance $G'_{th}(WK^{-1}m^{-2})$, whereas $C'_{th}(JK^{-1}m^{-2})$ represents the heat capacitance of the chip and the package. The heat convection provided by an air flow over the surface of the package is represented by $G'_{conv}(WK^{-1}m^{-2})$. As can be seen from figure 2.2

---

[1] $'$ indicates a quantity per unit area.

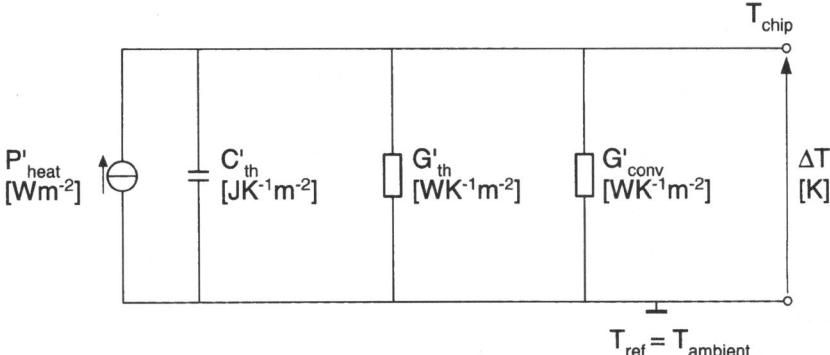

Figure 2.2. Network equivalent of the first order thermal model of a chip and its environment.

the temperature of the chip, $T_{chip}(K)$, is determined relative to the reference ambient temperature, $T_{ambient}(K)$, by the heat production of the internal circuits and the thermal conductivity of the chip and its package to the ambient. In case the maximum chip temperature will be exceeded extra conduction via the package can be provided by convection, e.g. generated by a fan. To determine if this could be the case one has to derive local chip temperatures from the local power densities.

The average electrical power $\overline{P}_{el}(W)$ consumed in a certain time interval equals:

$$\overline{P}_{el} = \frac{1}{\Delta t} \int_{t_0}^{t_0+\Delta t} P_{el}(t)dt \tag{2.1}$$

$P_{el}(W)$ is the instantaneous electrical power, see figure 2.1. The average electrical power consumed per unit area for a certain area of the chip is called electrical power density and will be indicated by $\overline{P'}_{el}(Wm^{-2})$. Suppose that this electrical power density is eventually completely converted into the heat per unit area $\overline{P'}_{heat}(Wm^{-2})$. Moreover, each part of the chip area can be modelled with the electrical equivalent circuit of figure 2.2. Therefore, the local chip temperature, which can be determined via the response of this circuit, equals:

$$T_{chip}(t) = T_{ambient} + \frac{\overline{P'}_{heat}}{G'_{th} + G'_{conv}} \left[1 - exp\left(-\frac{(G'_{th} + G'_{conv})t}{C'_{th}} \right) \right] \tag{2.2}$$

Temperature management problems caused by increasing power densities can be solved partially by using new developed packages with larger heat conductivities, equation 2.2. In extreme cases heat convection could be applied for extra heat transfer from the chip surface to the ambient. These mechanical solutions force up chip prices, a very undesirable development especially for consumer products demanding cheap, i.e. mostly plastic, packages. However, even the best and most expensive packages can not drain the amount of heat predicted by the Semiconductor Industry Association (SIA) for near future technologies. Therefore, other power reduction methods have to be found and applied to prevent heat from being generated in the first place.

2.2   Energy considerations

Mobile equipment like laptops, personal digital assistants (PDA) and cellular phones have to drain their power from a battery containing a limited amount of energy. It is evident that all these applications possess standby *modes*; the standby mode may be subdivided into standby[2] and active *periods*, see figure (2.3), like for instance in cellular phones. During the active periods the phone is partially activated and among other things checks for incoming calls. In the standby period the phone is inactive, only some necessary internal circuit states are retained. An important figure of merit of this type of mobile equipment is endurance of the battery pack in terms of maximum standby and active times:

$$T_{standby,max} = \frac{E_{battery}}{\overline{P}_{standby}}$$

$$T_{active,max} = \frac{E_{battery}}{\overline{P}_{active}}$$

(2.3)

$E_{battery}$ is the amount of energy stored in the battery, mostly indicated in amperes times hour (Ah), whereas $\overline{P}_{standby}$ is the average electrical power over all standby modes and $\overline{P}_{active}$ is the average power over all active modes. The times of equation 2.3 are figures of merit and are not only important from a practical point of view, they are also key marketing means used in fighting the competition. The average standby power drained during one standby mode, i.e. one standby time, equals the sum of the energy consumption during the active and standby period divided by the standby time. When considering the usefulness of the application of standby modes, the time aspect becomes important; application is only useful when the energy consumed during the standby mode is much less than during the active mode. For cellular phone applications the standby energy can only be kept low when the energy consumed during standby periods is much less than during active periods. For given average power levels in active and standby periods it therefore is of the utmost importance that for future cellular applications hardware and protocols are developed, which provide for as large as possible a ratio between those period times. In general this holds for the ratio between active and standby mode times.

One has to realize that given the capacity of the battery, its lifetime, equation 2.4, depends on the average power consumed during both the active and standby

---

[2]In literature standby modes (periods) are often called sleep modes (periods).

modes. Therefore, reduction of the active power is also an important issue to increase battery lifetime.

$$T_{lifetime} = \frac{E_{battery}}{\overline{P}_{total}} \tag{2.4}$$

$\overline{P}_{total}$ is the power averaged over all active and standby modes.

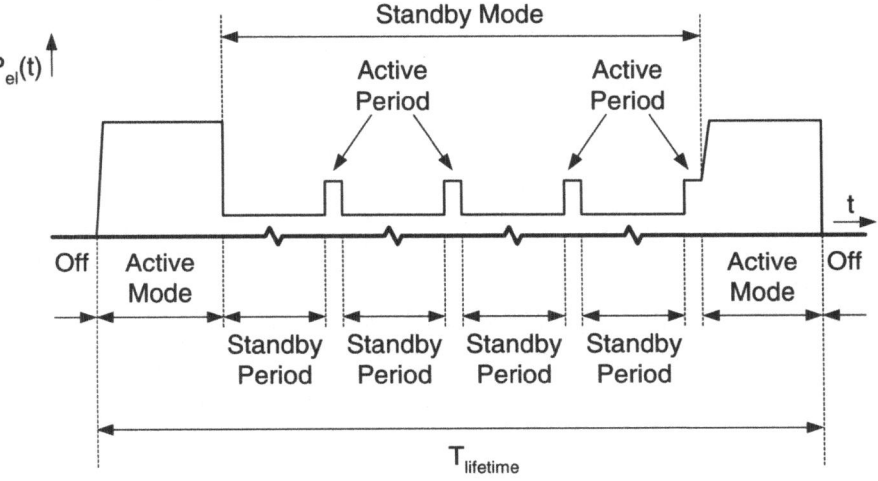

Figure 2.3. Operation modes of a cellular phone.

2.3 Conclusions

New CMOS technology generations are characterized by decreasing feature sizes, increasing clock frequencies and transistor densities causing both increased power consumption and power density. Increased power consumption in general drives up operational costs and especially for battery operated systems burdens the environment. Ever increasing power densities push (local) chip temperatures to physical limits.

For mobile applications in general, standby modes are only useful to extend battery lifetime when the energies consumed during these standby modes are much less than during the active modes. This is where the aspect of time comes in. For given average power levels in active and standby modes, it is important that hardware and protocols are developed, which provide for large ratios between standby and active times.

Since battery lifetime depends on the average power consumption during active and standby modes, reduction of both the active power and the standby power should be considered.

3

POWER DISSIPATION IN DIGITAL
CMOS CIRCUITS

From chapter 2 it has become clear that power dissipation plays an important role when considering operational costs, environment, local chip temperatures and battery lifetime. The power dissipation of a computational process depends on the physical process on which it is based. Two types of physical processes can be distinguished, see figure 3.1:

- reversible processes;

- irreversible processes.

Reversible in essence means that the order of actions can be reversed to return to the initial situation. A physical reversible process distinguishes itself from an irreversible process by the fact that all states can be traversed in opposite direction in such a way that returning to the initial state causes no changes in the system or its surroundings. This means that a reversible process dissipates no power and does not increase the entropy of the universe ($dS = 0$) in contrast to an irreversible process. This topic will be explained in section 3.1.

Reversible logic consists of logic ports, which return to their initial state, after performing logical operations, via the reverse sequence of states. No information in the system is destroyed, because it is required to return to the initial state, otherwise the return path is lost. When these logical operations are physically being performed quasi-statically[1], i.e. infinitely slowly, the process will be physically reversible. Practical reversible logic however, will have to perform logical operations in a limited amount of time, i.e. non quasi-statically.

---

[1] A quasi-static process is defined as an infinitely slowly evolving process, which is in thermal equilibrium at all times. However, this is physically impossible since slowly evolving and thermal equilibrium is a contradictio in terminis. Therefore, a slowly varying practical physical process will always be an approximation of a quasi-static process.

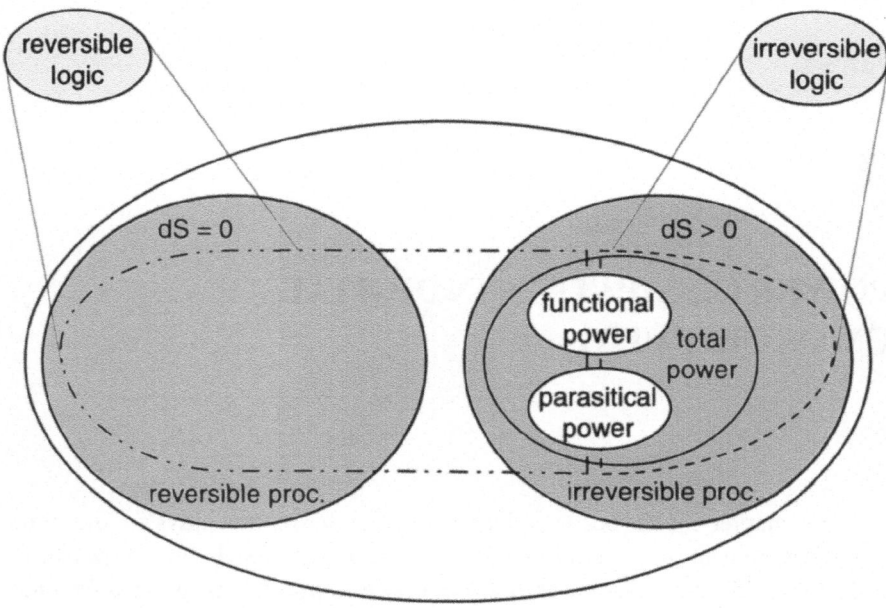

Figure 3.1. Subdivision of physical processes.

Although the power dissipation can be reduced considerably, they will dissipate power and therefore become physically irreversible, see figure 3.1.

Irreversible logic does not traverse the state sequences in the reverse order to return to the initial states after performing logical operations. Therefore, information is lost continuously causing power dissipation and entropy generation ($dS > 0$). This phenomenon is explained in section 3.1. Irreversible logic always is based on physical irreversible processes. For an irreversible process the theoretical minimal energy loss per logic bit will be determined in section 3.1.

The majority of present-day computational systems are produced in CMOS technology. In these systems all logic blocks consist of irreversible logic, hence they are also physically irreversible. Moreover, the power dissipation of present and (near) future practical systems is on the order of one billion times the fundamental limit. For this reason the power dissipation sources in deep submicron CMOS circuits will be considered. For each source the key parameters will be indicated. The total power dissipation of a digital CMOS circuit will be divided into two classes, see figure 3.1:

- functional power;

- parasitical power.

The *functional power* is the power needed to just change the internal states, i.e. the state-capacitors' charges, of a digital CMOS circuit on behalf of information processing[2]. The functional power dissipation will be discussed in section 3.2.

The *parasitical power* is power which is dissipated and dominant when the circuit is idle, defined as leakage power, or power which could be dissipated during state transitions without attributing to the actual changes of the internal states, defined as short-circuit power. The parasitical power dissipation, will be discussed in section 3.3.

The progress of CMOS technology reflects upon power dissipation. Power dissipation trends for both functional and parasitical power dissipation will be discussed in section 3.4.

3.1 Thermodynamics of computation

Considering the energetics of computing the first fundamental question to be posed is: what is the minimum energy required to carry out a computation? This minimum depends on the type of physical process the logic is based on. Two types of physical processes can be distinguished:

- reversible process;

- irreversible process.

In a *reversible process* all states can be traversed in opposite direction in such a way that returning to the initial state causes no changes in the system or its surroundings, i.e. in every cycle all the energy delivered to the process by an energy source is retrieved afterwards. Reversible processes are necessarily quasi-static, i.e. these processes move infinitely slowly through equilibrium states, and exhibit neither hysteresis nor friction.

An *irreversible process* moves through non-equilibrium states, i.e. is non quasi-static, before coming to a final equilibrium state. The path between two states is affected by transients, causing the process to deviate from the ideal path, thereby dissipating extra heat. Therefore, more paths can be followed to reach the same final equilibrium state. Once in the final state the information about which path has been followed, i.e. information about the speed by which the trajectory was traversed, is lost, preventing the process from following the same path backwards when the process is reversed. Extra energy will have to be supplied to the process in order to let it return to its initial state.

From these descriptions it can be concluded that reversibility is only possible in a quasi-static, i.e. infinitely slowly evolving, process and that, at least in theory, this process consumes no energy. Therefore, the work done *by* a process is in a non quasi-static way always less than in a quasi-static way, whereas

---

[2]The functional power is defined over one complete transition cycle of a circuit node (section 3.2).

the work done *on* a process is in a non quasi-static way always more than in a quasi-static way. In section 3.1.1 the fundamental limits of computation are determined for reversible and irreversible processes. Section 3.1.2 discusses the first and second laws of thermodynamics and provides background information on the concept of entropy. This discussion is used to explain that adiabatic and lossless are not synonym.

3.1.1   Fundamental limits of power dissipation

Consider the cylinder and frictionless pistons of figure 3.2 representing an equivalent physical model of the output stage of an electronic logic port. The

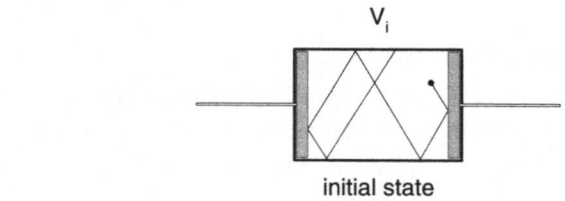

initial state

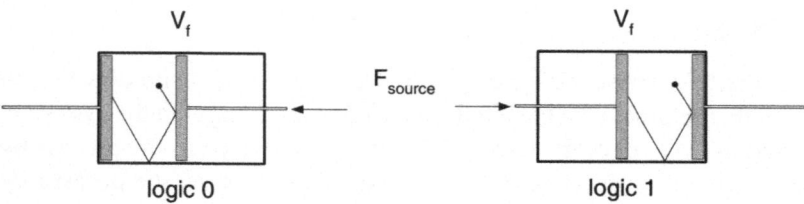

logic 0   logic 1

Figure 3.2.   Equivalent physical model of the output stage of an electronic logic port.

cylinder contains only one gas atom [15] of an ideal gas. In case the atom is in the left half of the cylinder, it counts as a 0 bit, if it's in the right half, it's a 1 bit. The upper half of figure 3.2 shows the initial situation, because the atom can move around freely in the whole volume of the cylinder, which is equivalent to the situation before power up of the logic port. During computation or information processing the output port will finally be forced into one of the two possible states, 0 or 1. In our example this means that one of the pistons has to shrink the volume in order to force the atom in the left or the right half of the cylinder, as indicated in the lower half of figure 3.2.

Suppose that the volume is shrunk isothermally, i.e. the whole system is immersed in a thermal bath at a fixed temperature T, and instantly, i.e. non quasi-statically, from V_i to V_f. How much work W will the energy source have to deliver to the piston? The work ΔW done on a system by an external pressure P_{ext} when the volume of the system changes by ΔV is given by:

$$\Delta W = P_{ext}\Delta V \tag{3.1}$$

This is true for any process, whether quasi-static or not, because P_{ext} is not a state variable of the system. Note that the work done on the system is negative in this definition, because ΔV is negative in this case. For an ideal gas it holds:

$$PV = n_a kT \qquad (3.2)$$

In our case the number of atoms n_a equals one, and concepts like temperature T, pressure P and volume V still make sense in this case as long as they are considered to be time averaged, smoothing out the irregularities of this one particle; k represents Boltzmann's constant. Since the initial and final temperatures and number of atoms are equal, the products of external pressure and volume in the initial and final state are equal:

$$
\begin{aligned}
P_i V_i &= P_f(V_i + \frac{W}{P_f}) \Longleftrightarrow \\
\Delta W &= V_i(P_i - P_f)
\end{aligned}
\qquad (3.3)
$$

P_i and P_f are the initial and final external pressure respectively, exerted by the energy source on the piston. The internal pressure in the cylinder becomes equal to the external pressure after thermal equilibrium has been reached.

Now suppose that the pressure on the one-atom gas is increased isothermally again, but in two steps in stead of one. In the first step the pressure is increased from P_i to P_h, with $P_i < P_h < P_f$, and in the second step from P_h to P_f. The work ΔW_1 done on the gas in the first step equals:

$$\Delta W_1 = V_i(P_i - P_h) \qquad (3.4)$$

In the second step the work ΔW_2 done on the gas equals:

$$\Delta W_2 = V_h(P_h - P_f) \qquad (3.5)$$

V_h is the volume at the end of the first step. The total work done on the gas is:

$$\Delta W_1 + \Delta W_2 = V_i(P_i - P_f) + (P_h - P_f)(V_h - V_i) \qquad (3.6)$$

Since $P_i < P_h < P_f$ and $V_i > V_h > V_f$, the additional term $(P_h - P_f)(V_h - V_i)$ in equation 3.6 is positive and so the work done on the gas is smaller when the pressure is increased in two steps compared to the case in which it is increased in one step. This argument can be continued, carrying out the pressure increase in more and more steps, reducing the total amount of work done on the gas. The *minimum* amount of work is done when the pressure is increased gradually, proceeding in infinitesimal steps, i.e. quasi-statically. The work takes place quasi-statically only if the difference between the external applied pressure P_{ext} remains arbitrarily close to the internal pressure of the system P. For a quasi-static process, the work can be calculated in terms of *system state variables* only:

$$dW = PdV \qquad (3.7)$$

W is not a state variable; therefore, the differential, called diminutive, is denoted by $\bar{d}W$. From the discussion above it has become clear that the work done on the gas is minimal when the work is done quasi-statically, therefore:

$$\bar{d}W_{nonquasi-static} = P_{ext}dV < \bar{d}W_{quasi-static} = PdV \qquad (3.8)$$

Now the amount of work done on the one-atom ideal gas during the quasi-static isothermal compression is discussed. This process is in principle reversible, because it is quasi-static. To do this, as has been stated before, the whole system is immersed in a thermal bath at a fixed temperature T such that at all times the system is in thermal equilibrium with its surroundings. The piston is moved by the energy source exerting force F_{source}, decreasing the volume from V_i to V_f, see figure 3.2. Combining equation 3.1 and 3.2 the amount of work equals:

$$\Delta W = \int_{V_i}^{V_f} \frac{n_a kT}{V} dV = n_a kT \ln \frac{V_f}{V_i} \qquad (3.9)$$

In our case $n_a = 1$ and the final volume V_f is half the initial volume V_i, therefore the minimal work done on the one-atom gas to force the output in a defined state equals:

$$\Delta W_{minimal} = -kT \ln 2 \qquad (3.10)$$

The work done on the gas is converted into internal gas heat and drained off into the thermal bath. The physical state of the atom before and after isothermal compression is the same, i.e. its internal (kinetic) energy has not changed, $dE_{int} = 0$. However, the possible spatial positions the atom can occupy, i.e. the number of configurations, has been decreased, resulting in an increase of the information about its configuration. In thermodynamics the number of configurations is expressed by the quantity entropy S. The less configurations are available in the system, the lower the entropy. In case the decrease of the number of configurations increases the ordering, the information increases. Hence, information is related to ordering and not necessarily to the number of configurations[3]. In this case of isothermal quasi-static compression ($dE_{int} = 0$), the relation between the minimum amount of work done on the gas and the decrease in entropy is as follows, see equation 3.15 in section 3.1.2:

$$\Delta W_{minimal} = T\Delta S_{minimal} \implies \Delta S_{minimal} = -k \ln 2 \qquad (3.11)$$

During the evaluation of the state of the atom the energy source maintains its force on the piston. No work has to be done to do this. The state of this logic

---

[3]Sometimes the increase of entropy is explained as the increase of disorder in a system. However, the entropy can even increase when the ordering increases. In this case an increase in ordering provides for an increase in the number of possible configurations, e.g. in the spontaneous isolated process of (liquid) crystal formation.

port can be observed by some lossless process which triggers the next port in line to force its output in a defined state. At the end of the calculation, the force on the piston can be removed, enabling the gas to expand isothermally, thereby draining $kT \ln 2$ joule from the heat bath and converting it into work on the piston. This work will be restored in the energy source. Because no energy was lost during the whole cycle, the entropy of the system and its environment did not change, which is a property of reversible processes.

In order to be able to follow the same state sequence, i.e. path, backwards as forwards an intermediate logic port can only be brought back into its initial state when its inputs are still defined [16, 17]. When running a computer forward, there must be no ambiguity in the forward step. With a reversible machine, there can be no ambiguity in backward steps either; logical reversibility is a necessary precondition for physical reversibility.

In case the work delivered by the gas, during the isothermal expansion, is not restored in the energy source, the process would be irreversible and $kT \ln 2$ joule of energy [18] would be lost for every bit of information that is being destroyed in this way. Erasure of information leads to heat flow into the universe, thereby increasing the entropy of it. This was discovered by Charles Bennett in 1961 [16, 17], thereby solving the problem of Maxwell's Demon [15, 19]. Maxwell's Demon describes the separation of hot and cold particles in a box by a daemon, thereby creating a temperature difference between both sides of the box. The box and the demon form an isolated system. This means that the entropy of the system has decreased, in clear violation with the second law of thermodynamics, which states that the entropy of an isolated system remains equal or increases. The mathematical proof of the second law is presented in section 3.1.2. In the early days it was believed that a minimum energy of $kT \ln 2$ was needed to determine whether a particle is hot or cold. Bennett however, stated that the demon, after putting a particle in the right part of the box, destroys information by selecting the next particle, because he can only remember the state of one particle. Therefore, after the selection of a particle the state of the previous particle has been forgotten, making it impossible to put the previous particle back where it came from. The latter is a necessary precondition for reversibility. So the destruction of information leads to dissipation and increases the entropy of the universe.

3.1.2 Thermodynamic laws

The *first law* of thermodynamics is based on the law of conservation of energy and defines heat. The first law of thermodynamics in differential form reads [20]:

$$dE_{int} = đQ_h - đW \tag{3.12}$$

dE_{int} represents the change in the internal energy of the process. $\text{d}W$ is the change in the work and $\text{d}Q_h$ is the change in the heat supplied. For a reversible process it holds that for every amount of heat $\text{d}Q_h$ entering the system the change in entropy dS equals:

$$dS = \frac{\text{d}Q_h}{T} \tag{3.13}$$

Substitution of equations 3.7 and 3.13 in equation 3.12 gives the *fundamental thermodynamic relation*:

$$dE_{int} = TdS - PdV \tag{3.14}$$

This relation will be used to place a lower bound on the entropy change when an amount of heat $\text{d}Q_h$ enters a system. It already has been shown that the amount of work done on a system is minimal for a reversible process, equation 3.8. Therefore, for any process it holds that:

$$\text{d}W \leq \text{d}W_{reversible} = PdV = TdS - dE_{int} \tag{3.15}$$

The *second law* of thermodynamics indicates in which direction a process, which obeys the first law of thermodynamics, evolves spontaneously. Substitution of equation 3.12 in equation 3.15 gives the mathematical form of the second law of thermodynamics:

$$\text{d}Q_h \leq TdS \tag{3.16}$$

The equality holds for reversible processes, for any real process the change is greater. This shows that when an amount of heat $\text{d}Q_h$ is added to a system, the entropy of the system rises at least $\text{d}Q_h/T$.

In literature the term adiabatic is often used as a synonym for lossless, for instance in adiabatic logic or adiabatic switching. However, adiabatic means that there is no heat exchange between the process and its surroundings, i.e. $\text{d}Q_h = 0$. This has nothing to do with whether the process is reversible or not. As discussed before, physical processes are lossless when they are reversible and vice versa. Processes can only be reversible (lossless) when they are quasi-static, and there is no friction and hysteresis involved. Since the temperature T is non-zero, the entropy of a system which undergoes a reversible, adiabatic process remains unchanged, and is therefore lossless:

$$\text{d}Q_h = TdS = 0 \Longleftrightarrow dS = 0 \tag{3.17}$$

From the second law of thermodynamics, equation 3.16, it follows that the entropy of a system which undergoes an irreversible, adiabatic process is always positive, since:

$$\text{d}Q_h = 0 \land T > 0 \Longrightarrow dS > 0 \tag{3.18}$$

Despite the fact that this is an adiabatic process the change in entropy dS is positive, therefore the process is lossy and irreversible. Thus physically reversible can be a synonym for lossless.

3.2 Functional power dissipation

The functional power (figure 3.3) is dissipated to just attribute to state changes, i.e. the state-capacitors' charges, of a digital CMOS circuit in favor of logic operations. At present power reduction is an important issue, not only for economic and environmental reasons, but also for practical reasons, because local chip temperatures are pushed to the limit, as has been discussed in chapter 2. A general expression for the functional power dissipation facilitates a systematic approach for finding suitable power reduction techniques, discussed in chapter 4, dealing with the components of this expression. In order to derive an expres-

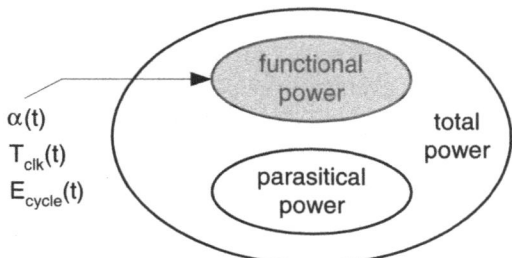

Figure 3.3. Parameters influencing the functional power dissipation.

sion for the functional power dissipation, a node of a system with the following properties will be considered:

- variable clock period T_{clk};

- variable node transition-cycle activity α;

- variable switching energy E_{cycle}.

The clock period can change as a function of parameter variations [9], for instance temperature fluctuations, or as a function of the workload [10], see figure 3.4. The clock period of interval i equals:

$$T_{clk,i} = t_i - t_{i-1} \qquad (3.19)$$

The node transition-cycle activity (α) is the number of up and down transitions a circuit node traverses in one period of the clock, i.e. one cycle consists of one up and one down transition. This number can differ per node and period. The switching energy is the energy delivered by or to the power supply source during switching transitions. This energy depends on the capacitance of the node, the supply voltage level and the way the node is charged or discharged.

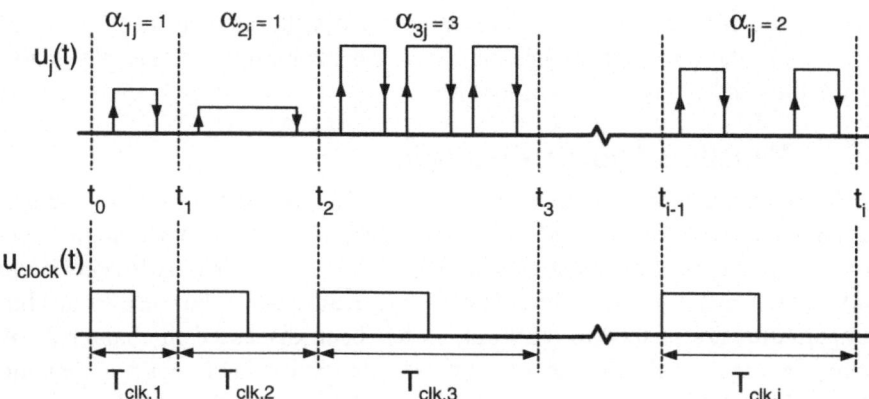

Figure 3.4.   System with variable node transition-cycle activity, clock period and switching energies.

The average functional power is the functional power averaged over all nodes and clock periods; first the average over all nodes within one clock period is determined, thereafter this is averaged over all clock periods. The average functional power within a clock period ($T_{clk,i}$) is determined by taking the sum of the average functional power per node of all nodes (N) in the circuit. The average functional power per node on its turn will be divided in a sum of the average power for up and down transitions.

The average functional power of all circuit nodes (N) within a clock period ($T_{clk,i}$) equals:

$$\overline{P}_{func,i} = \frac{1}{T_{clk,i}} \sum_{j=1}^{N} \int_{t_{i-1}}^{t_i} p_{ij}(t)dt \qquad (3.20)$$

$p_{ij}(t)$ represents the power delivered by or to the power supply, in clock period i, to charge or discharge node j, and N represents the total number of nodes in the circuit. Figure 3.5, which is a magnification of the clock period $T_{clk,i}$ of figure 3.4, shows an arbitrary power function $p_{ij}(t)$ for the up and down transitions of node j. The energy delivered by the power supply in this clock period equals:

$$E_{func,i} = \overline{P}_{func,i} T_{clk,i} \qquad (3.21)$$

Suppose that for node j all up transitions are equal and all down transitions are equal. Under worst case conditions, i.e. the power function becomes zero before the next transition, the total energy of α cycles equals α times the energy of one cycle. Therefore, the power function $p_{ij}(t)$ is redefined as the sum of two separate power functions $p_{up,ij}(t)$ and $p_{dn,ij}(t)$. Figure 3.6 shows the redefined power functions. The power delivered by the power supply in clock period i to charge node j equals $p_{up,ij}(t)$, whereas $p_{dn,ij}(t)$ represents the power delivered

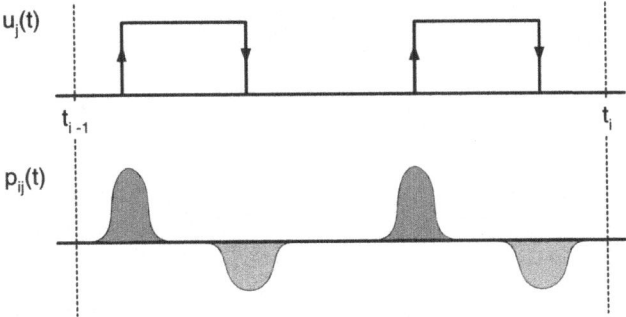

Figure 3.5. Power delivered by or to the power supply during switching transitions of node j.

to the power supply in clock period i to discharge node j. In this case equation 3.21 becomes:

$$E_{func,i} = \sum_{j=1}^{N} \alpha_{ij} \int_{t_{i-1}}^{t_i} \left[p_{up,ij}(t) + p_{dn,ij}(t) \right] dt \qquad (3.22)$$

$$= \sum_{j=1}^{N} \alpha_{ij} E_{cycle,ij}$$

α_{ij} represents the node cycle activity of node j in clock period i, whereas $E_{cycle,ij}$ represents the net dissipated energy delivered by the power supply to node j in clock period i. For reversible node transition-cycles, discussed in

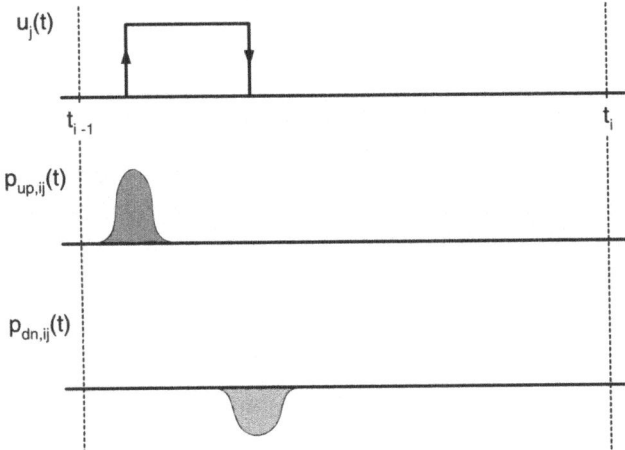

Figure 3.6. Power functions for node transitions.

section 4.3, the sum of $\int p_{up,ij}(t)dt$ and $\int p_{dn,ij}(t)dt$ equals zero, because no energy is lost to the environment. For resistive node transition cycles $p_{dn,ij}(t)$ equals zero, because all the stored energy is dissipated.

The overall average functional power dissipation is equal to the sum of the energy per interval, equation 3.22, divided by the sum of all clock period times:

$$\overline{P}_{func} = \lim_{n \to \infty} \frac{1}{\sum_{k=1}^{n} T_{clk,k}} \sum_{i=1}^{n} E_{func,i} \tag{3.23}$$

$$= \lim_{n \to \infty} \frac{1}{\sum_{k=1}^{n} T_{clk,k}} \sum_{i=1}^{n} \sum_{j=1}^{N} \alpha_{ij} E_{cycle,ij}$$

Signal skews can cause extra glitching node transitions, i.e. increase α. Since signal skews depend on the switching energy via the supply voltage U_{dd}, there exists a dependency between α and E_{cycle}. However, it is assumed here that α and E_{cycle} are independent random variables. Under these conditions equation 3.23 becomes:

$$\overline{P}_{func} = \lim_{n \to \infty} \frac{n}{\sum_{k=1}^{n} T_{clk,k}} N \overline{\alpha} \overline{E}_{cycle} \tag{3.24}$$

$$= \frac{N}{\overline{T}_{clk}} \overline{\alpha} \overline{E}_{cycle}$$

An equivalent frequency is defined as the frequency with which a circuit has to be clocked to consume the same amount of power as with different clock frequencies in separate time intervals:

$$f_{clk,eq} \stackrel{\Delta}{=} \frac{1}{\overline{T}_{clk}} \tag{3.25}$$

Substitution of equation 3.25 in equation 3.24 gives:

$$\overline{P}_{func} = N \overline{\alpha} f_{clk,eq} \overline{E}_{cycle} \tag{3.26}$$

In chapter 4 techniques will be discussed, which reduce the functional power dissipation via the parameters α, f_{clk} and E_{cycle}. Equation 3.26 transforms into the average functional power for *one* circuit node, a form often found in literature, when either one resistive up or down transition is considered:

$$\overline{P}_{func,edge} = \frac{1}{2} \alpha_{lit} f_{clk} C_{node} U_{dd}^2 \tag{3.27}$$

With α_{lit} representing the number of node transitions in one clock cycle in stead of the number of transition cycles, and C_{node} representing the capacitance of a circuit node.

3.3   Parasitical power dissipation

In section 3.1.1 it has been argued that the minimum power dissipation for an irreversible logical operation equals $kT \ln 2$. However, the expression for the functional power dissipation, derived in the previous section, indicates that the amount of power dissipated in present-day logical systems does not even come close to the theoretical limit. Therefore, although the functional power is dissipated in favor of executing a logical operation, all excess power dissipation is a waste. All the worse is power dissipation when a circuit is idle or power dissipation without attribution to changes of circuit states: parasitical power dissipation. The parasitical power dissipation will be divided into two main groups (figure 3.7), depending on whether the leakage is device or circuit related:

- leakage power dissipation;

- short-circuit power dissipation.

The *leakage power dissipation* is device related parasitical power, which is dissipated even when the circuit is idle. It becomes dominant in standby modes, since no information is being processed and therefore the functional and the short-circuit powers are zero. For every new MOS technology, channel lengths and oxide thicknesses decrease and threshold voltages drop, resulting in increasing leakage power. All the parameters contributing to the leakage power will be discussed in section 3.3.1. With the help of these parameters suitable power reduction techniques, discussed in chapter 5, counteracting these leakage mechanisms can be found.

Short-circuit power dissipation is caused by rail-to-rail currents during state transitions, without attributing to the actual changes of the internal states, and is only apparent in static CMOS logic circuits. In contrast to the leakage power it is circuit related parasitical power. The short-circuit power dissipation will be discussed in section 3.3.2.

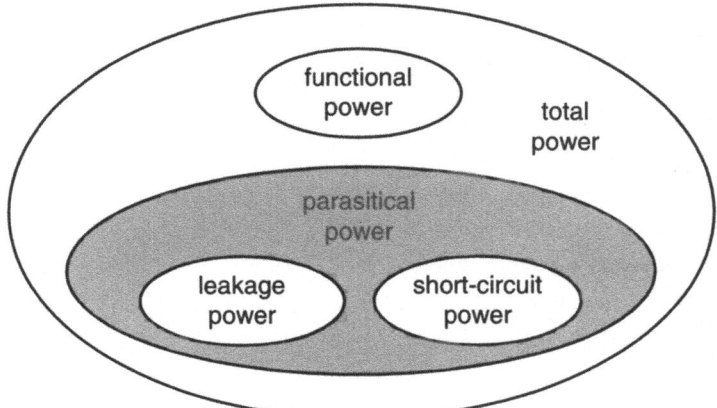

Figure 3.7. Subdivision of the parasitical power dissipation in CMOS circuits.

All formulas and examples in this section will be defined for n-channel MOS transistors. Analogous expressions can be found for PMOS devices.

3.3.1   Leakage power dissipation

The leakage power will be divided into three sub-groups distinguished by their origins (figure 3.8):

- channel leakage current;
- diode leakage current;
- gate leakage current.

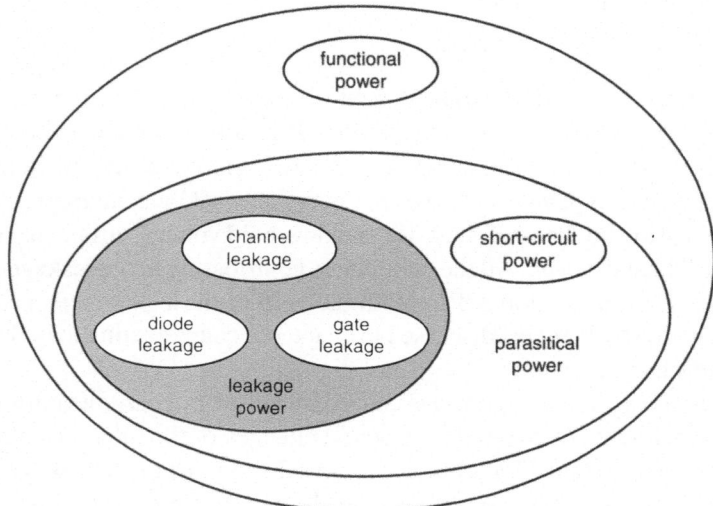

Figure 3.8.   Subdivision of the leakage power dissipation.

The *channel leakage current*, indicated by I_1 in figure 3.9, consists of drain-source currents which are present even when gate-source voltages are zero. The *diode leakage current*, indicated by I_2, consists of source and drain pn-junction leakage currents. The *gate leakage current*, indicated by I_3, consists of direct tunneling current components. These leakage mechanisms are discussed in subsections 3.3.1.1 through 3.3.1.3.

3.3.1.1   Channel leakage current

Three effects contributing to the channel leakage current (I_1 in figure 3.9) will be distinguished:

- weak-inversion current;
- drain-induced barrier lowering (DIBL) current;
- channel edge current.

These effects are discussed in the next paragraphs.

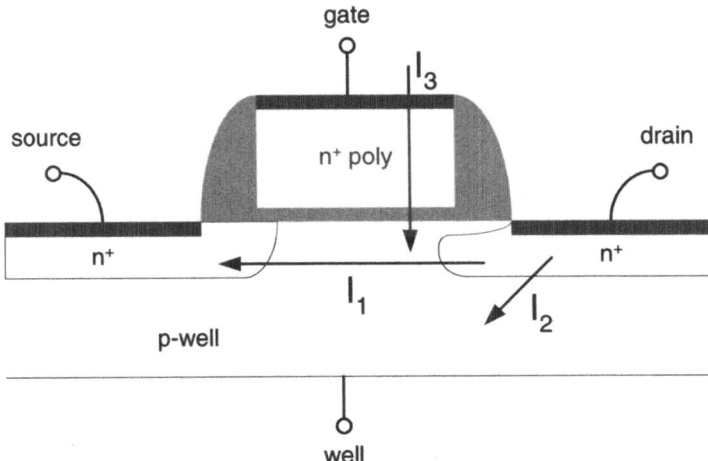

Figure 3.9. Overview of the leakage current mechanisms in deep sub-micron MOS transistors.

Weak-inversion current. In literature very often the terms sub-threshold and weak inversion are lumped together. However, according to more precise definitions [21] the threshold voltage U_{th} lies just outside the weak-inversion region, i.e. in the moderate-inversion region, see figure 3.10. The equations, which are only valid in the weak-inversion region are being used for the whole region below the threshold voltage, thereby completely ignoring the moderate-inversion region. In this book the equations for the weak-inversion conditions will only be used for the weak-inversion region. The definition for the threshold voltage is according to the one generally accepted in literature, see equation 3.38. The definitions of the weak-inversion currents, in [21], will be expressed as a function of this threshold voltage, as is commonplace in literature. Therefore, the weak-inversion condition is defined as follows:

$$U_{ds} \geq 0 \qquad (3.28)$$
$$U_L \leq U_{gs} < U_M$$

U_{ds} is the drain-source potential, whereas U_{gs} is the gate-source potential. The boundary potentials U_L (figure 3.10), which is the transition voltage between depletion and weak inversion, U_M, which is the transition voltage between weak and moderate inversion, and U_H, which is the transition voltage between moderate and strong inversion, can be expressed as follows [21]:

$$U_L = U_{fb} + \phi_F + \gamma\sqrt{\phi_F + U_{sb}} \qquad (3.29)$$

$$U_M = U_{fb} + 2\phi_F + \gamma\sqrt{2\phi_F + U_{sb}} \qquad (3.30)$$

$$U_H = U_M + U_Z \qquad (3.31)$$

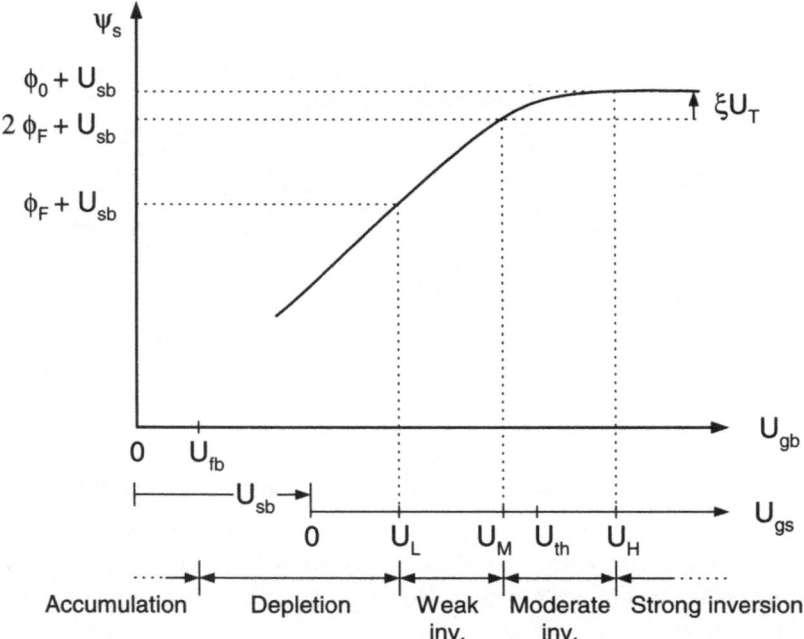

Figure 3.10.   Operation regions of an MOS transistor as a function of the surface potential ψ_s
[21].

U_{fb} represents the flat-band voltage, i.e. the gate voltage at which the valence
and conduction bands are not bent, ϕ_F indicates the Fermi potential, and U_{sb} is
the source-bulk potential. The value of U_Z is several tenths of a volt [21]. For
the flat-band voltage holds:

$$U_{fb} = \phi_{gc} - \frac{Q_{ss}}{C_{ox}} \tag{3.32}$$

in which Q_{ss}, C_{ox} and ϕ_{gc} represent the fixed charge in the gate oxide, the gate
capacitance and the work function difference between the gate and the channel
material, respectively. At the onset of weak inversion the surface potential
ψ_s equals $\phi_F + U_{sb}$, whereas at the onset of moderate inversion ψ_s equals
$2\phi_F + U_{sb}$. For the body factor γ holds:

$$\gamma = \frac{\sqrt{2qK_{Si}N_{sub}}}{C'_{ox}} \tag{3.33}$$

N_{sub}, C'_{ox}, q and K_{Si} represent the substrate doping concentration, the gate
capacitance per unit area, the electron charge and the permittivity of silicon,
respectively. In the weak-inversion region carriers, injected at the source end of
the channel, move by diffusion to the drain, similar to charge transport across the

base of bipolar transistors. In this regime the drain-source current I_{ds} changes exponentially with the gate-source voltage. The weak-inversion current for long-channel NMOS transistors equals [21]:

$$I_{ds} = I_M \exp\left(\frac{U_{gs} - U_M}{nU_T}\right)\left(1 - \exp\left(\frac{-U_{ds}}{U_T}\right)\right) \qquad (3.34)$$

in which U_T is the thermal or Boltzmann voltage:

$$U_T = \frac{kT}{q} \qquad (3.35)$$

and n indicates the slope factor:

$$\begin{aligned} n &= 1 + \frac{C_d'}{C_{ox}'} \qquad &(3.36)\\ &= 1 + \frac{\gamma}{2\sqrt{2\phi_F + U_{sb}}} \end{aligned}$$

in which C_d' indicates the depletion capacitance per unit area. I_M can be written as follows:

$$\begin{aligned} I_M &= \mu_0 C_{ox}' \frac{\gamma}{2\sqrt{2\phi_F + U_{sb}}} U_T^2 \frac{W_{ch}}{L_{ch}} \qquad &(3.37)\\ &= \mu_0 \frac{\sqrt{2qK_{Si}N_{sub}}}{2\sqrt{2\phi_F + U_{sb}}} U_T^2 \frac{W_{ch}}{L_{ch}} \end{aligned}$$

in which W_{ch} and L_{ch} indicate the channel width and length, respectively, whereas μ_0 represents the zero bias mobility. Note that I_M is independent of C_{ox}'.

Equation 3.34 can also be expressed as a function of the threshold voltage U_{th}. This can be accomplished by writing U_M as a function of the threshold voltage. The threshold voltage will be defined in terms of the strong-inversion surface potential [21]:

$$\begin{aligned} U_{th} &= U_{th0} + \Delta U_{th} \qquad &(3.38)\\ &= U_{fb} + \phi_0 + \gamma\sqrt{\phi_0 + U_{sb}} \\ &= U_{th0} + \gamma\left(\sqrt{\phi_0 + U_{sb}} - \sqrt{\phi_0}\right) \end{aligned}$$

The surface potential in strong inversion equals [21]:

$$\phi_0 + U_{sb} = 2\phi_F + \xi U_T + U_{sb} \qquad (3.39)$$

For uniformly doped substrates ξ equals approximately 6 [21]. Substitution of equation 3.38 in equation 3.30 yields:

$$
\begin{aligned}
U_M &= U_{th} - \xi U_T - \gamma \left(\sqrt{2\phi_F + \xi U_T + U_{sb}} - \sqrt{2\phi_F + U_{sb}} \right) \quad (3.40) \\
&\approx U_{th} - \xi U_T \left(1 + \frac{\gamma}{2\sqrt{2\phi_F + U_{sb}}} \right)
\end{aligned}
$$

In equation 3.40 the difference has been approximated by a differential[4]. Substitution of equation 3.36 in equation 3.40 yields:

$$
U_M \approx U_{th} - \xi n U_T \tag{3.41}
$$

Since U_M is now expressed in terms of the threshold voltage, substitution in the weak-inversion current of equation 3.34 gives:

$$
I_{ds} = I_0 \exp \left(\frac{U_{gs} - U_{th}}{n U_T} \right) \left(1 - \exp \left(\frac{-U_{ds}}{U_T} \right) \right) \tag{3.42}
$$

Moreover, for I_0 it holds:

$$
I_0 = I_M \exp \xi \tag{3.43}
$$

It should be noted that I_{ds} in equation 3.42 depends on U_{sb} via the quantities U_{th}, I_0 and n.

As an example, the weak-inversion region for a 0.25μm transistor is depicted in figure 3.11 as the linear portion of the curve.

For constant U_{ds} the weak-inversion current is influenced by five factors:

- gate-source voltage;

- threshold voltage;

- weak-inversion "slope";

- device geometries.

- temperature.

All these factors will be discussed successively.

The exponential relationship between the weak-inversion current and the *gate-source potential* is used deliberately for analog applications, for instance to solve logarithmic and anti-logarithmic calculations, or differential equations

---

[4]
$$
\sqrt{2\phi_F + \xi U_T + U_{sb}} - \sqrt{2\phi_F + U_{sb}} \approx \xi U_T \frac{d}{d(2\phi_F + U_{sb})} \left(\sqrt{2\phi_F + U_{sb}} \right)
$$

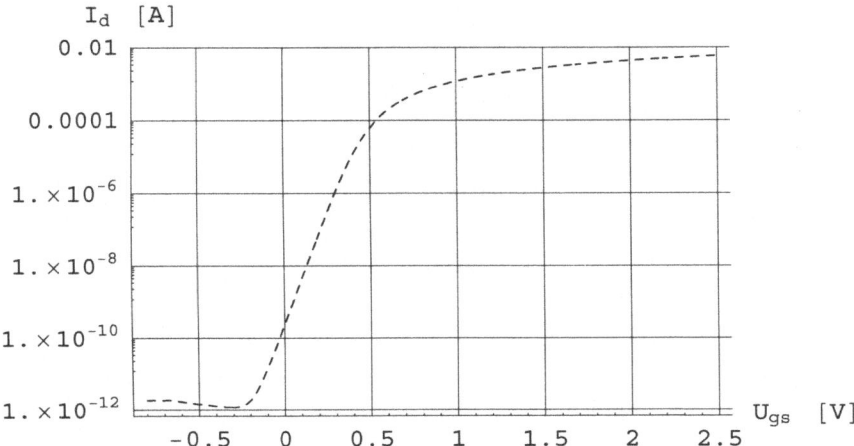

Figure 3.11.   The $\log(I_d)$ versus U_{gs} curves of a $0.25\mu m$ transistor, for $U_{ds} = 2.5$ V.

in dynamic trans-linear circuit applications. However, in digital applications a weak-inversion current is just leakage, which should be reduced to the minimum. For positive gate-source potentials weak-inversion currents increase, whereas for negative ones they decrease (equation 3.42). However, for too negative potentials channel currents increase again, see figure 3.11, caused by leakage from drain to bulk and leakage through the gate oxide. These effects will be discussed in section 3.3.1.2 and 3.3.1.3, respectively.

An increasing *threshold voltage* is responsible for an exponentially decreasing weak-inversion current. The data-processing speed performance also decreases for an increasing threshold voltage, since it is proportional to $(U_{dd} - U_{th})^\sigma$. U_{dd} represents the supply voltage and σ, which lies between 1 and 2, mainly indicates the effect of carrier saturation. Carrier saturation drives σ towards 1. According to the required level of leakage current on the one hand, and speed performance on the other hand an optimal threshold voltage can be chosen.

The *weak-inversion "slope"*, S_{wi}, is the inverse of the slope of the linear portion of the curve in figure 3.11 and indicates the change in U_{gs} necessary to change the weak-inversion current one decade. The weak-inversion slope, sometimes called sub-threshold slope, can be deduced from equation 3.42 and is expressed as:

$$S_{wi} = \frac{dU_{gs}}{d\ln(I_{ds})} ln10 = nU_T \ln 10 \qquad (3.44)$$

The minimal (ideal) weak-inversion slope at room temperature equals:

$$\lim_{C'_d/C'_{ox} \to 0} S_{wi} = U_T \ln 10 = 60 \ mV/decade \qquad (3.45)$$

A weak-inversion slope greater than 100 mV/decade is an indication that the device is leaky. A lower weak-inversion slope indicates a lower off-state leakage current, i.e. the weak-inversion current at $U_{gs} = 0$, for a given threshold voltage.

Devices geometries are scaled down in every new technology generation to increase the functionality per unit chip area. The effect scaling has on the drain-source current can be accounted for by replacing the threshold voltage U_{th} in the long-channel equation 3.42 by an effective threshold voltage U_{th}^*.

$$U_{th}^* = U_{th} - \Delta U_{th} \qquad (3.46)$$

ΔU_{th} accounts for the threshold voltage shift caused by variations in the channel length and width of a transistor.

For a short-channel transistor the effective threshold voltage will drop compared to a long-channel transistor. Emanating from a long-channel transistor the effect of the decreasing channel length is that the same gate charge per unit area controls less depletion charge per unit area in the region under the channel of a short-channel transistor. This can be explained as follows. The depletion width around the drain is larger than the depletion width under the channel, due to the higher voltage of the drain. Bringing the source and drain closer together will increase the influence of the drain on the channel and the source, and therefore the depletion width around the source and under the channel increase. Hence, the depletion capacitance C_d' decreases from the source along the channel to the drain. For the same gate-bulk potential the surface potential increases, from the source, along the channel, to the drain. Hence, more carriers will be attracted into the channel, which increases the drain-source current for a given gate-source potential. This increase in drain-source current can be modelled as a lower effective threshold voltage, called U_{th} roll-off, see figure 3.12. This is one of the possible ways to explain short-channel effects on the effective threshold voltage; more can be found in [21].

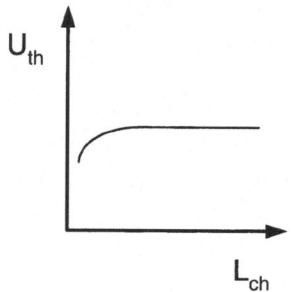

Figure 3.12.   Effective threshold voltage as a function of the channel length L_{ch}.

Decreasing the channel *width* influences the effective threshold voltage differently for different isolation technologies. For isolation technologies with less abrupt transitions between the channel and the isolation area, e.g. LOCal

Oxidation of Silicon (LOCOS) [22] and Polysilicon Buffered LOCOS (PBL) [23], the effective threshold voltage will increase for narrow channels [24], see figure 3.13a. In literature this is often referred to as the narrow-width effect. Because of the less abrupt transition the depletion region extends partly under

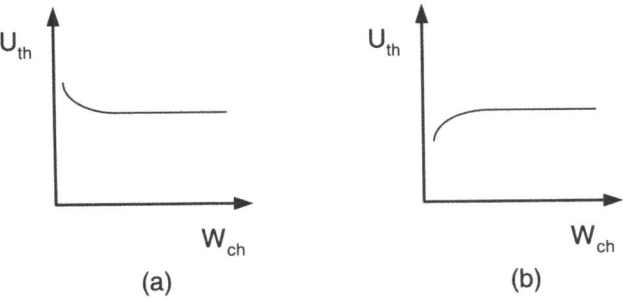

(a)   (b)

Figure 3.13. Effective threshold voltage as a function of the channel width W_{ch}. (a) For LOCOS and PBL, (b) for SILO and STI.

the field oxide, see figure 3.14. Especially for narrow channels the amount of depletion charge under the field oxide becomes relatively large compared to the depletion charge under the channel. Since the gate has to deplete a larger area, more gate charge, i.e. a larger gate voltage, will be necessary to attract the same amount of carriers into the channel. For isolation technologies containing

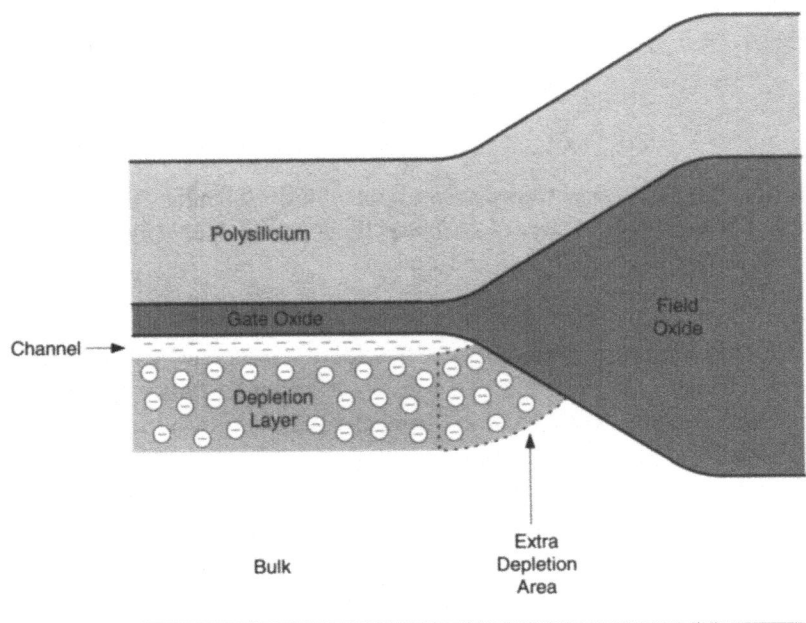

Figure 3.14. LOCOS field oxide.

more abrupt, e.g. Sealed Interface Local Oxidation (SILO) [25], and abrupt,
e.g. Shallow Trench Isolation (STI) [26], transitions between the channel and
the isolation area, the effective threshold voltage decreases for narrow channels
[24, 27, 28], see figure 3.13b. In literature this is often referred to as the inverse
narrow-width effect and the theory behind it will be discussed in the paragraph
dealing with the channel edge current on page 35.

Temperature affects the weak-inversion current through the following parameters:

- carrier mobility μ;

- Boltzmann voltage U_T;

- slope factor n;

- threshold voltage U_{th};

The temperature dependence of the carrier mobility can be expressed as
follows [21, 29]:

$$\mu(T) = \mu(T_r) \left(\frac{T}{T_r}\right)^{-\nu} \tag{3.47}$$

with ν between 1.2 and 2 and T_r represents room temperature. The Boltzmann
voltage, presented in equation 3.35, varies linearly with temperature. The slope
factor is temperature dependent via the Fermi potential ϕ_F, as can be seen from
equation 3.36. For the Fermi potential holds:

$$\phi_F = U_T \ln\left(\frac{N_{sub}}{n_i}\right) \tag{3.48}$$

The intrinsic carrier concentration n_i is exponentially dependent on temperature
[21, 30]. The threshold voltage U_{th} reduces linearly with increasing temperature
[21]:

$$U_{th}(T) = U_{th}(T_r) - \chi(T - T_r) \tag{3.49}$$

in which χ is the threshold voltage temperature coefficient, which is usually
between $0.5 \ mV \cdot K^{-1}$ and $3 \ mV \cdot K^{-1}$, with larger values in this range
corresponding to heavier doped substrates, thicker oxides, and smaller source-
bulk potentials.

Therefore, for higher temperatures the weak-inversion slope increases, i.e.
less steep, and the threshold voltage decreases, leading to an exponentially
increasing weak-inversion current for increased temperatures [29].

Drain-induced barrier lowering current.   The drain electrode, like the gate
electrode, also influences the channel current. In long-channel transistors an

increase in the drain potential mainly widens the depletion region around the drain. The wider the local depletion width t_d, see figure 3.15, underneath the channel, the smaller the local depletion capacitance C'_d. For a constant gate-bulk potential U_{gb} this effects in a local increase in the surface potential ψ_s, i.e. a stronger bending of the energy bands [30] (figure 3.15), since the ratio C'_d/C'_{ox} decreases. A higher local surface potential attracts more carriers into the channel, although its impact depends on the distance from the source.

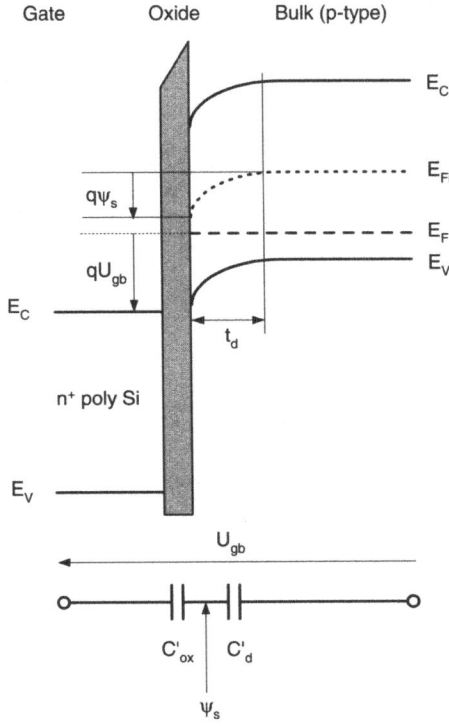

Figure 3.15. Energy-band diagram through an NMOS transistor with n^+ polysilicon gate.

In a long-channel transistor only a small part of the depletion region under the channel, in the vicinity of the drain, is influenced by its potential. Since the drain is located relatively far from the source, the width of the depletion region and the surface potential near the source are not influenced. Consequently, the potential barrier between the source and the channel is not lowered, figure 3.16a.

In short-channel transistors, however, the drain is located so close to the source that it even interacts with the depletion region around the source. In this situation the width of the depletion region underneath the whole channel is influenced by the drain potential. Hence, the drain is able to increase the surface potential at the source side of the channel, i.e. the potential barrier is lowered, making it much easier for the source to inject carriers into the channel, for a given gate potential, figure 3.16b. This phenomenon is called Drain-Induced Barrier Lowering (DIBL). Since DIBL enhances the carrier concentration in the channel, it increases the off-state current. Equation 3.42 is extended for DIBL in equation 3.50 by the factor ϑ [14]:

$$I_{ds} = I_0 \exp\left(\frac{U_{gs} - U_{th} + \vartheta U_{ds}}{nU_T}\right)\left(1 - \exp\left(\frac{-U_{ds}}{U_T}\right)\right) \qquad (3.50)$$

DIBL moves up the $I_{ds} - U_{gs}$ characteristics for higher drain voltages, as can

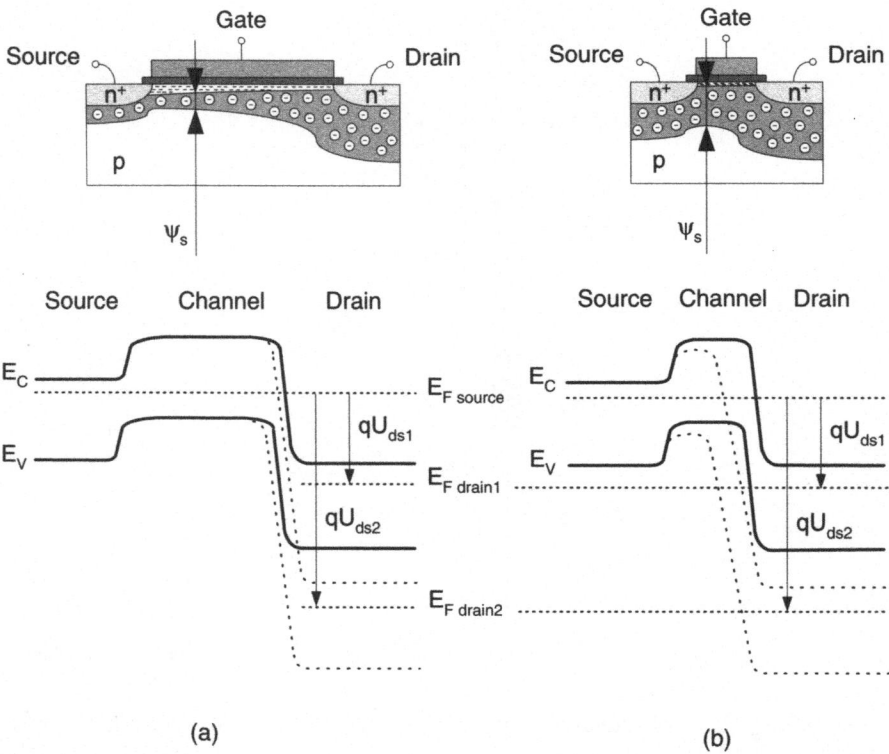

(a)   (b)

Figure 3.16. The effects of DIBL on (a) long-channel and (b) short-channel NMOS transistors and their respective equipotential plots.

be seen in figure 3.17.

Excessive high drain-source voltages may lead to punch-through, which is an extreme form of DIBL. Since punch-through is a degeneration leakage effect appearing outside the normal range of operation it will not be discussed in detail. In the case of punch-through the gate loses control and a large current exists deep in the sub-gate region [30, 31]. The punch-through current varies quadratically

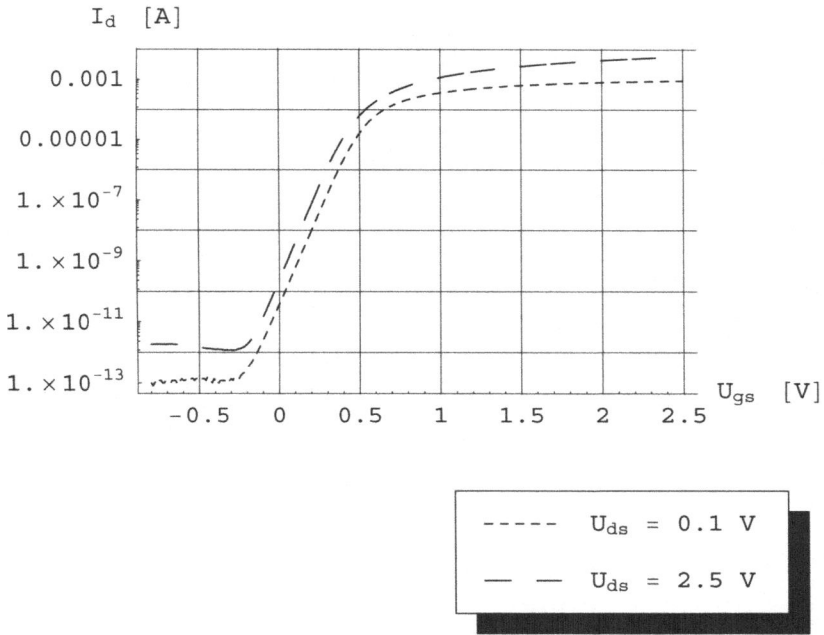

Figure 3.17. The $\log(I_d)$ versus U_{gs} curves of a 0.25μm transistor, for $U_{ds} = 0.1$ V and $U_{ds} = 2.5$ V (upper curve), showing DIBL.

with the drain-source voltage and the weak-inversion slope increases, reflecting the increase in drain leakage [32].

Channel edge current. Technological developments to boost integration densities not only focus on shrinking device geometries, but also on reducing transistor isolation areas. Until the 0.25μm CMOS technology generation different types of oxide growth based technologies have been used, evolving toward an increasingly abrupt transition between the active and the isolation area: LOcal Oxidation of Silicon (LOCOS)[22] down to 0.7μm, Polysilicon Buffered LOCOS (PBL) [23] for 0.5μm and Sealed Interface Local Oxidation (SILO) [25] for 0.35μm. This evolution culminated in Shallow Trench Isolation (STI) [26], which accommodates much smaller design dimensions and is used since the 0.25μm CMOS technology. In the STI technology the chemical oxide, i.e. tetraethylorthosilicate (TEOS) is deposited in a trench, allowing for a completely abrupt transition. An abrupt transition eliminates the lateral encroachment of the field oxide into the channel area, called bird's beak, a side effect of LOCOS. Abrupt transitions can cause the weak-inversion current near the channel edges to increase considerably, which can be visualized as a parasitic transistor with a lower effective threshold voltage parallel to the intrinsic channel transistor. The impact the isolation technique has on the edge current depends on the following parameters, which are also indicated in figure 3.18:

- the isolation height h with respect to the silicon surface;

- the transition angle θ_t;

- the corner radius r;

- the oxide fixed charge density N_{ss};

- the channel doping concentration N_{sub};

- the oxide thickness t_{ox}.

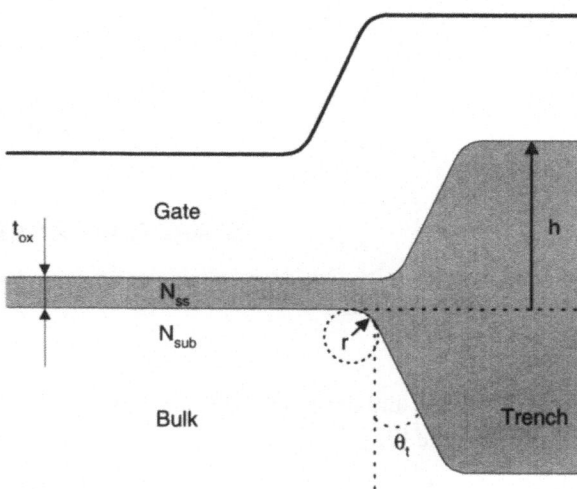

Figure 3.18. Schematic representation of an abrupt transition from the active to the isolation area.

The height of the isolation oxide h at the edge of the channel is the most important parameter, because it influences the control of the gate on the channel edge. To observe the influence of the fabrication process on h, the oxide growth will be examined. The growth of the gate oxide takes place after the trench oxide deposition. Before the gate oxide can be grown, the trench oxide has to be removed from the channel region by a local deoxidation step. Depending on the etching technology used, this may lead to a recessed, i.e. buried, trench oxide near the channel edge. The recessed oxide is indicated schematically in the Local Deoxidation area of figure 3.19. Figure 3.20 shows a tunneling electron microscope (TEM) cross section image of an STI-isolated transistor. A recessed trench oxide, i.e. negative h, induces a charge sharing area (figure 3.19) at the edge of the channel caused by gate fringing fields. In this situation the gate attracts more carriers in the channel edge area, giving rise to increased edge leakage currents, i.e. a reduced effective threshold voltage. For STI-isolated transistors with narrow-width channels, the side edges overlap and therefore the threshold voltage reduces compared to the one expected for a wide channel transistor. This narrow channel effect is just opposite to the effects known in oxide growth technologies. It has been reported in [24] that the difference in

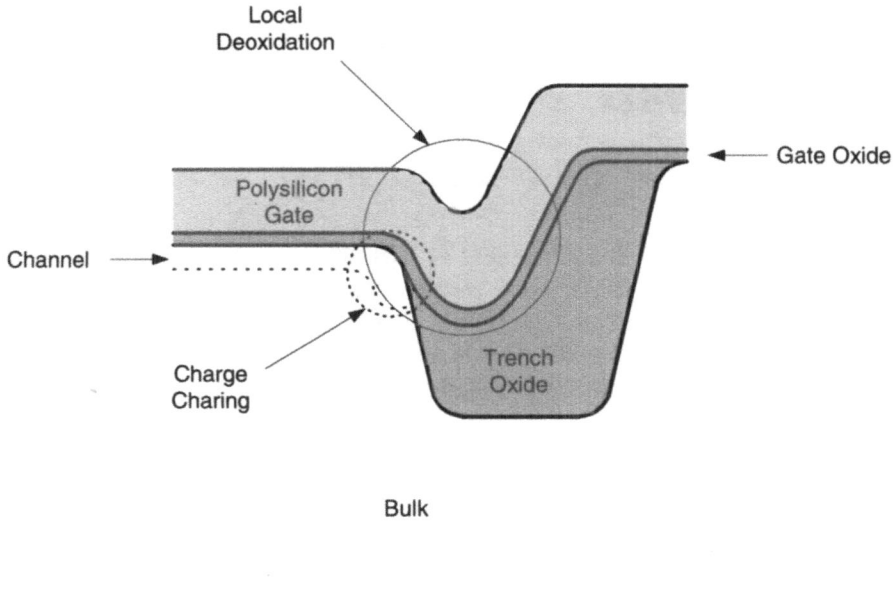

Figure 3.19. Schematic cross section of an STI-isolated transistor.

weak-inversion current between an overflow oxide ($h = +35nm$) and a buried oxide ($h = -35nm$) equals 2 decades, for a transition angle $\theta_t = 0$. The case of an overflow oxide ($h = +35nm$), results in an almost non-existent parasitic edge transistor.

By inclining the transition by an angle θ_t of $20°$ the parasitic edge current is reduced by one decade and by rounding the corner of the edge by a radius of curvature r of 100nm, the weak-inversion current is reduced by half a decade.

The quality of the oxide with respect to the fixed charge density (N_{ss}) has opposing effects on the NMOS and PMOS transistors. Since these fixed charges always have a positive nature, they reduce the effective threshold of the edge transistors for NMOS and increase it for PMOS transistors.

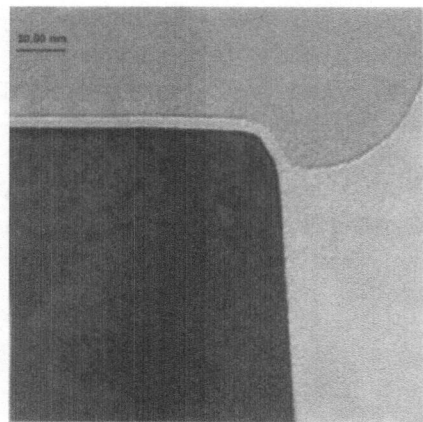 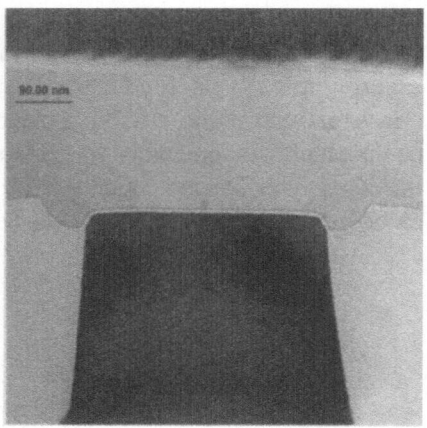

Figure 3.20.   TEM cross section of an STI-isolated transistor.

Defects are located along the edges as well. The combination of high electrical fields and traps cause trap-assisted tunneling, which increase channel edge currents. Decreasing the gate oxide thickness t_{ox} and the channel doping concentration N_{sub} reduces the drop in effective threshold voltage of the edge transistor, compared to the main channel transistor. This is a consequence of the gate fringing field at the channel edge, which increases the ratio of the gate capacitance C'_{ox} and C'_d at the edge [28]. Since the gate has more control on the channel edge than on the main channel, the parasitic edge transistors are less sensitive to short channel effects and substrate bias effects. Therefore, biasing the substrate reduces the channel weak-inversion current more aggressively compared to the edge weak-inversion current. The resulting weak-inversion hump, see figure 3.21, indicates that the edge current becomes the dominant leakage current. In [33] a method is described to identify the hump effect. Figure 3.22 shows a 50μ wide and 0.25μ long transistor with 20 edges showing no weak-inversion hump, despite the substrate biasing. For this transistor STI leakage currents are not dominant.

3.3.1.2   Diode leakage current

The diode leakage current (I_2 in figure 3.9) can be divided into two components:

- reverse bias leakage;

- gate-induced drain leakage (GIDL).

These components are discussed in next two paragraphs.

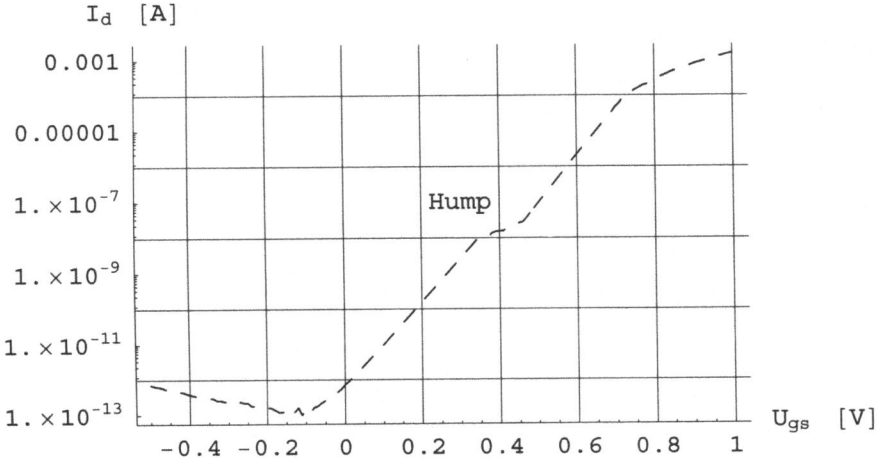

Figure 3.21. The weak-inversion hump revealing edge leakage currents.

Reverse bias leakage. Source-bulk and drain-bulk pn-junctions are reverse biased during normal operation of an MOS transistor. The resulting reverse-bias leakage current can be divided into two parts:

- reverse saturation current;

- generation current.

The reverse saturation current is the fundamental reverse-bias current in a pn-junction. The generation current is the pn-junction current produced by the thermal generation of electron-hole pairs within the space charge region. The total reverse-bias current equals:

$$I_{reverse-bias} = A_j \left(J_s + J_{gen} \right) \qquad (3.51)$$

in which A_j is the pn-junction area, J_s is the reverse saturation current density, and J_{gen} is the generation current density [30]. The reverse saturation current is described by:

$$J_s = \left(\frac{q D_n n_{p0}}{L_n} + \frac{q D_p p_{n0}}{L_p} \right) \qquad (3.52)$$

in which D_n and D_p are the electron and hole diffusion coefficient, respectively. L_n and L_p are the diffusion lengths for electrons and holes, respectively, whereas n_{p0} and p_{n0} represent the thermal-equilibrium concentrations of minority carriers in a p-type region, i.e. electrons, and in an n-type region, i.e. holes, respectively [30]. The Einstein relation links the electron, D_n, and hole, D_p, diffusion coefficients with the electron, μ_n, and hole, μ_p, mobilities and

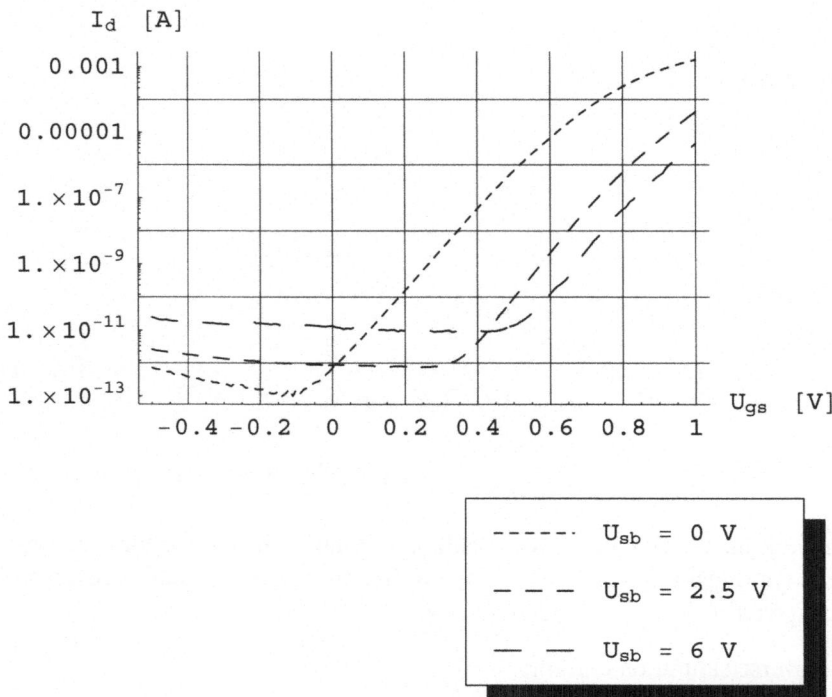

Figure 3.22. A 50μ wide and 0.25μ long transistor with 20 edges, without a weak-inversion hump, i.e. STI-leakage is not dominant.

the Boltzmann voltage U_T as follows [30]:

$$\frac{D_n}{\mu_n} = \frac{D_p}{\mu_p} = \frac{kT}{q} = U_T \tag{3.53}$$

For the thermal-equilibrium minority electron concentration in a p-type region, n_{p0}, and the thermal-equilibrium minority hole concentration in an n-type region, p_{n0}, hold [30]:

$$n_{p0} = \frac{n_i^2}{N_d} \tag{3.54}$$

$$p_{n0} = \frac{n_i^2}{N_a}$$

in which n_i represents the intrinsic concentration of electrons, and N_d and N_a indicate the donor and acceptor doping concentrations, which approximate the majority hole and electron concentration, respectively. Substitution of equations 3.53 and 3.54 in equation 3.52 yields:

$$J_s = q n_i^2 U_T \left(\frac{\mu_n}{L_n N_d} + \frac{\mu_p}{L_p N_a} \right) \tag{3.55}$$

The generation current consists of electrons and holes, which are thermally generated within the space charge region. The generation current is described by:

$$J_{gen} = \frac{q n_i t_{dep}}{2 \tau_l} \tag{3.56}$$

in which t_{dep} represents the depletion width, n_i the intrinsic concentration of electrons and τ_l the average carrier lifetime. The depletion width in, for instance, an NMOS $n^+ - p$ drain-bulk junction is voltage dependent and equals:

$$t_{dep} = \frac{2 K_{Si} \left(U_{bi} + U_{db} \right)}{q N_a} \tag{3.57}$$

in which U_{bi} and U_{db} represent the built-in potential barrier [30] and the drain-bulk potential, respectively, whereas N_a is the bulk acceptor doping concentration.

Gate-induced drain leakage. Gate-induced drain leakage (GIDL) [34] is a leakage current from drain to substrate, caused by high electric fields between gate and drain. Consider the region where the gate overlaps the n^+-drain, with the gate grounded and the drain region at U_{dd}, as shown in figure 3.23. A large

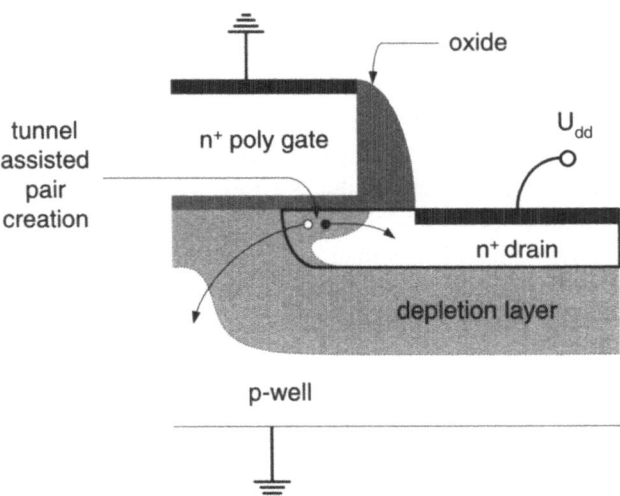

Figure 3.23. Gate-induced drain leakage (GIDL) in a schematic view of the gate-drain overlap region for a grounded gate and a drain biased at U_{dd}.

field E_{ox} exists in the oxide and a charge Q'_d per unit area is induced in the drain electrode:

$$Q'_d = K_{ox}E_{ox} \qquad (3.58)$$

in which K_{ox} represents the permittivity of the gate oxide. Charge Q'_d is provided by a depletion layer in the drain, which becomes a non-equilibrium deep-depletion layer for large E_{ox}, because minority carriers that might form an inversion layer are drained laterally to the substrate. The non-equilibrium surface region, which could collect minority carriers except for the lateral draining to the substrate, will be called the "incipient inversion layer" in the drain. For large enough E_{ox}, the voltage drop in the deep-depletion layer becomes large enough to allow tunneling in the drain via near-surface traps. In that case several trap-assisted events become possible. The trap-assisted events are typically present for low electric fields and are a strong function of temperature [35]. For field strengths above approximately $7MV/cm$ electron hole pairs will be generated through band-to-band tunneling. This generation current depends on the band gap E_g and is very sensitive to the electric field, see equation 3.59 [35],

$$I_{GIDL} \propto AE_{ox}^{\frac{5}{2}} \exp\left(-\frac{B}{E_{ox}}\right) \qquad (3.59)$$

where $A \propto E_g^{-\frac{7}{4}}$ and $B \propto E_g^{\frac{3}{2}}$ are constants.

The minority carriers emitted to the incipient layer are then laterally removed to the substrate, completing a path for a gate-induced drain current. Figure 3.24 shows the effect of GIDL on the drain current of a 0.25 μm NMOS transistor. The dotted line extension of the $U_{ds} = 2.5V$ curve indicates the source current, i.e. the drain current without GIDL.

3.3.1.3   Gate leakage current

Aggressive scaling of device geometries also affects the gate oxide thickness. The use of ultrathin gate oxides reduces short channel effects and enhances device performance, but opens the door wide to gate leakage currents (I_3 in figure 3.9). Gate oxides of less than 2 nm will be required for sub-100 nm generation MOS transistors, causing gate oxide tunneling. Two gate tunneling current components can be distinguished:

■ gate-to-channel direct tunneling current;

■ source and drain extension-to-gate overlap tunneling current.

Both tunneling current components can be either direct or trap-assisted. Trap-assisted tunneling is caused by defects in the insulator's crystal lattice or by encapsulated impurities and require less energy to cross the insulator.

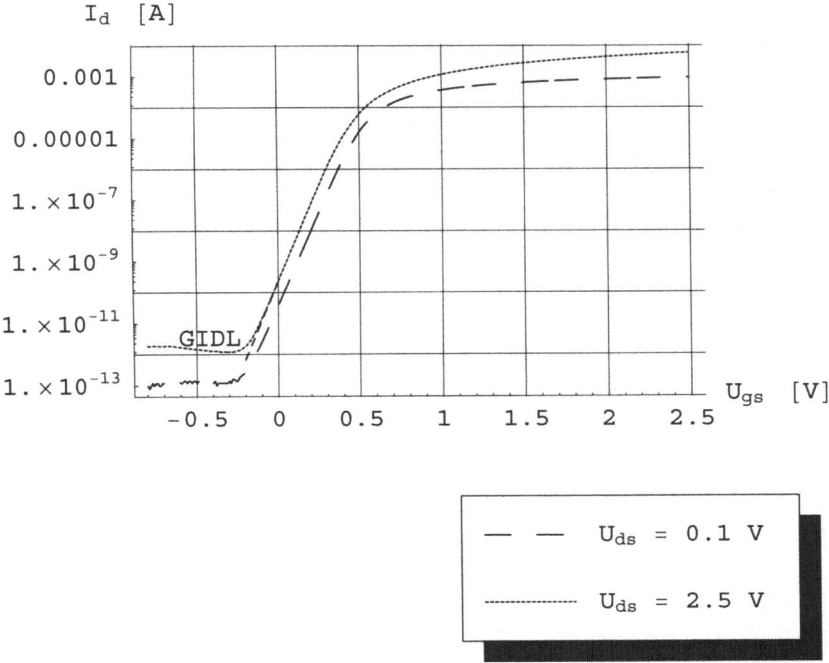

Figure 3.24. The effect of GIDL on the drain current (Id) of a $0.25\mu m$ NMOS transistor.

Gate-to-channel direct tunneling. Gate-to-channel direct tunneling becomes considerable, i.e. about 50 nA/μm^2 [36] for field strengths of 0.7 V/nm. These values come in reach for $0.13\mu m$ and smaller technologies with gate oxides of between 1 and 3 nm thick and supply voltages between 0.9 V and 1.8 V. For electric field strengths above 0.7 V/nm the tunneling current through the gate oxide will increase exponentially. This exponential tunneling mechanism is called *Fowler-Nordheim tunneling*. The Fowler-Nordheim current density J_{FN} (in A/cm^2) equals [31]:

$$J_{FN} = C_1 E_{ox}^2 \exp\left(-\frac{C_2}{E_{ox}}\right) \quad (3.60)$$

E_{ox} is the electrical field in the gate oxide (in MV/cm), C_1 (in A/MV^2) and C_2 (in MV/cm) are constants. In [36] models for direct tunneling currents in MOS devices can be found.

The gate-to-channel tunneling current I_{gc} in an n^+-polysilicon / SiO_2 / p-substrate NMOS structure, i.e. at the cross section from gate to channel (figure

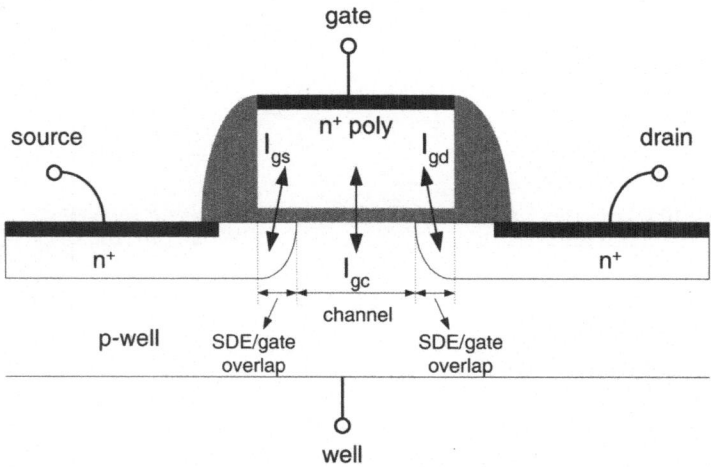

Figure 3.25.   Gate-to-channel and source/drain extension-to-gate overlap direct tunneling current components.

3.25), increases rapidly and becomes the dominant gate leakagecurrent when the gate voltage is below the channel flat-band voltage. In that case both the n^+-polysilicon gate and the p-type substrate are in accumulation, leading to lots of electrons available for tunneling through the gate oxide. Figure 3.26 shows the drain, gate and substrate currents as a function of the gate source voltage, for a 10μ wide, 0.3μ long NMOS transistor with a drain source bias voltage of $1.5V$ and an oxide thickness of $1.6nm$. At room temperature, a substrate doping concentration of $5 \cdot 10^{17}/cm^3$, and an oxide thickness of 1.5 nm, the channel flat-band voltage, $U_{fb,gc}$, is about -1 volt. For gate voltages from -1 to -1.5 volt the gate-to-channel current forms the dominant part of the drain current [37], marked by the arrow and I_{gc}. GIDL currents will become dominant for gate voltages below -2 volt, marked by the arrow and I_{GIDL}. This example indicates that for ultrathin oxide transistors gate leakage currents become dominant over GIDL for low negative gate bias values.

Source and drain extension (SDE) to gate overlap tunneling current. The source and drain extension (SDE) to gate overlap tunneling current I_{sg} and I_{dg} in an n^+-polysilicon / SiO_2 / n^+-drain MOS structure, i.e. at the cross section from gate to source or drain (figure 3.25), becomes the dominant gate leakage current for gate voltages between the channel flat-band voltage and the SDE flat-band voltage, $U_{fb,sde}$, [37] marked by the arrow and I_{dg}. The SDE flat-band voltage is approximately 0 volt, since both sides of the gate oxide are highly n^+ doped and become degenerate. Moreover, the voltage drop over the oxide can be neglected, because of the large C'_{ox} and therefore does not contribute to the SDE flat-band voltage. When the gate voltage is between 0 volt and the

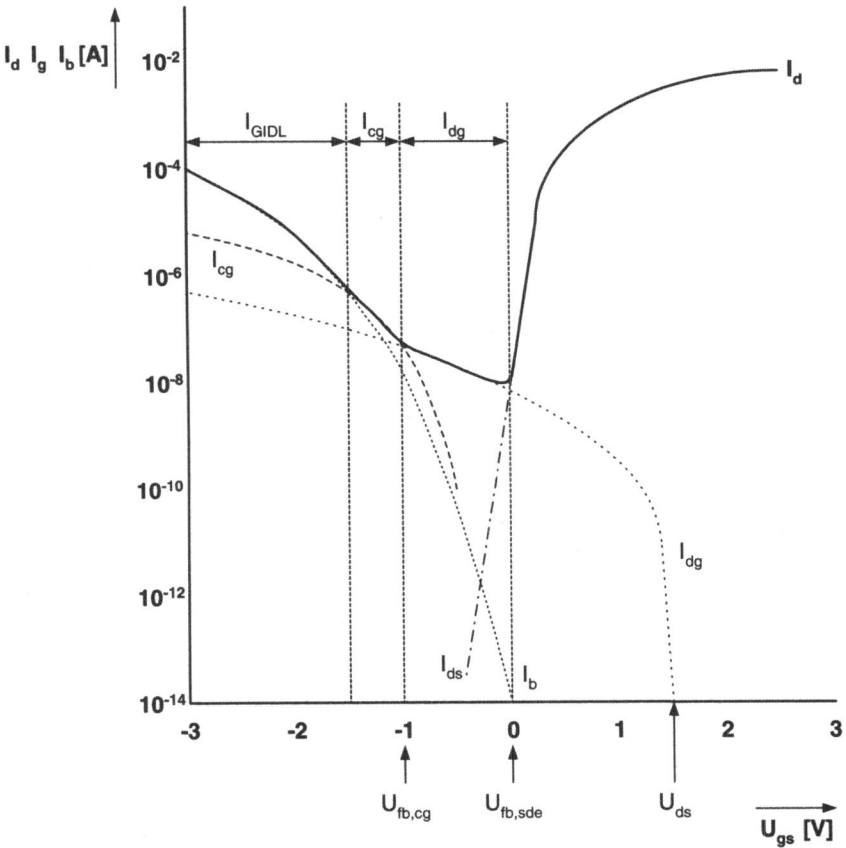

Figure 3.26. GIDL, gate-to-channel and SDE to gate overlap currents dominating the drain current for $U_{gs} < 0V$, for an NMOS transistor with $W_{ch} = 10\mu$m, $L_{ch} = 0.3\mu$m and $t_{ox} = 1.6nm$ [37].

channel flat-band voltage, the electrons in the n^+-polysilicon gate accumulate, leading to a non-negligible SDE-to-gate tunneling current.

3.3.2 Short-circuit power dissipation

Short-circuit power dissipation, like functional power dissipation, occurs during logic state transitions. Short-circuit power dissipation is circuit related parasitical power dissipation. In *static* CMOS logic finite rise and fall times of the input signals cause, during the short switching periods, direct paths between the power supply lines. Specifically, when the input signal voltage is in the range where both NMOS and PMOS transistors are in strong inversion, i.e. $U_{th,n} \leq U_{in} \leq U_{sw} - |U_{th,p}|$, a conductive path exists between the power supply lines, because both the NMOS and PMOS devices are conducting, see figure 3.27. $U_{th,n}$ and $U_{th,p}$ indicate the NMOS and PMOS threshold voltages, respectively, whereas U_{sw} represents the signal swing. Consider an input signal

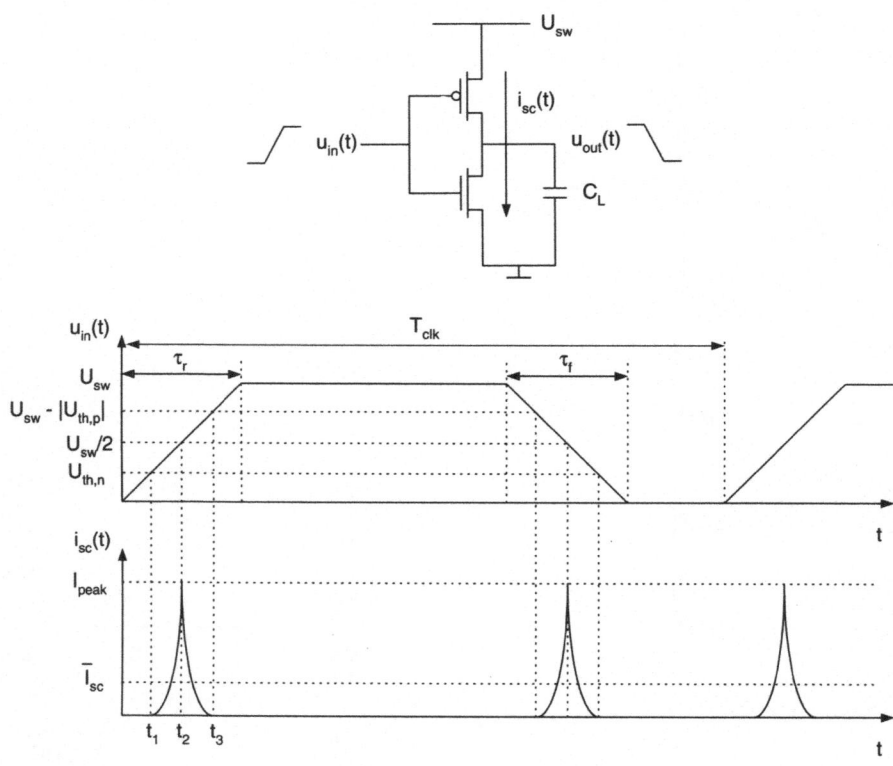

Figure 3.27. Short-circuit currents in a static CMOS inverter [38].

of an inverter with equal rise time τ_r and fall time τ_f. The average short-circuit current will be determined assuming no output load, equal threshold voltages U_{th} and equal effective transistor strengths β for NMOS and PMOS transistors [38]. In the interval from t_1 to t_2 the gate drain voltage of the NMOS transistor is smaller than U_{th}. Therefore, the NMOS transistor is in saturation. Since the PMOS is in saturation as well the short-circuit current equals:

$$i_{sc}(t) = \frac{\beta}{2}\left(u_{in}(t) - U_{th}\right)^{\sigma} \qquad (3.61)$$

in which $u_{in}(t)$ represents the input voltage. For $u_{in}(t)$, during the input ramp τ_r holds:

$$u_{in}(t) = \frac{U_{sw}}{\tau_r}t \qquad (3.62)$$

The current drive of the NMOS and PMOS parts are equally strong, therefore the short-circuit current will reach its maximum at $u_{in} = \frac{U_{sw}}{2}$. The average short-circuit current $\overline{I}_{sc}$ equals:

$$
\begin{aligned}
\overline{I}_{sc} &= \frac{2\alpha}{T_{clk}} \int_{t_1}^{t_3} i_{sc}(t)dt \\
&= \frac{4\alpha}{T_{clk}} \int_{t_1}^{t_2} \frac{\beta}{2} \left(u_{in}(t) - U_{th}\right)^\sigma dt
\end{aligned}
\tag{3.63}
$$

Due to the assumed symmetrical current drive the average current between t_1 and t_2 equals the average between t_2 and t_3. The node transition-cycle activity, α, represents the average number of transition cycles of a node, i.e. pairs of current peaks in figure 3.27, in a clock period T_{clk}, as defined in section 3.2. This parameter is part of the functional as well as the short-circuit power dissipation, because it is typical for power dissipated during logic transitions. The clock period is represented by T_{clk}. The value of σ lies between 1 and 2 depending on the intensity of carrier velocity saturation. Substitution of $u_{in}(t_1) = U_{th}$ and $u_{in}(t_2) = \frac{U_{sw}}{2}$ in equation 3.62, respectively yields:

$$
t_1 = \frac{U_{th}}{U_{sw}}\tau_r
\tag{3.64}
$$

$$
t_2 = \frac{\tau_r}{2}
$$

Substituting equation 3.64 in equation 3.63, and assuming that $U_{sw} \geq 2U_{th}$, delivers the average short-circuit current for α transition cycles per clock period:

$$
\begin{aligned}
\overline{I}_{sc} &= \frac{2\alpha\beta\tau_r}{T_{clk}U_{sw}} \int_{\frac{U_{th}}{U_{sw}}\tau_r}^{\frac{\tau_r}{2}} \left(\frac{U_{sw}}{\tau_r}t - U_{th}\right)^\sigma d\left(\frac{U_{sw}}{\tau_r}t - U_{th}\right) \\
&= \frac{2\alpha\beta\tau_r}{T_{clk}U_{sw}\,(\sigma+1)} \left(\frac{U_{sw}}{\tau_r}t - U_{th}\right)^{\sigma+1} \Bigg|_{\frac{U_{th}\tau_r}{U_{sw}}}^{\frac{\tau_r}{2}}
\end{aligned}
\tag{3.65}
$$

which results in:

$$
\overline{I}_{sc} = \alpha f_{clk}\frac{\beta\tau_r}{2^\sigma U_{sw}\,(\sigma+1)} \left(U_{sw} - 2U_{th}\right)^{\sigma+1}
\tag{3.66}
$$

The average short-circuit power dissipation for this inverter equals:

$$
P_{short-circuit} = \overline{I}_{sc}U_{sw} = \alpha f_{clk}\frac{\beta\tau_r}{2^\sigma\,(\sigma+1)} \left(U_{sw} - 2U_{th}\right)^{\sigma+1}
\tag{3.67}
$$

The trend in short-circuit power dissipation is regarded for constant field scaling. Constant field scaling scales U_{sw} and U_{th} with $1/\kappa = 0.7$, whereas β scales with κ. The clock frequency doubles for every generation. For a bench mark algorithm α is constant. Assuming the rise time τ_r also scales with $1/\kappa$, the short-circuit power per transistor scales between $2(1/\kappa)^3$ ($\sigma = 2$, best case) and $2(1/\kappa)^2$ ($\sigma = 1$, worst case). Since transistor and die areas scale with $(1/\kappa)^2$ and 1.14, respectively, the number of transistors increases by approximately a factor of $\kappa^2 \times 1.14 = 2.3$. Therefore, for $1/\kappa = 0.7$ the short-circuit power per die increases worst case with $2\,(1/\kappa)^2 \times 2.3 \approx 2.3$ for every new generation.

3.4   Trends in power dissipation

The SIA road map [1] describes CMOS technology trends, which leading chip manufacturers agreed upon could be achieved in due time. The tremendous drive behind the rapid progress of CMOS technology is the cry for increased speed performance and functionality from business and consumer markets. To increase speed performance and functionality per unit area and time it is necessary to increase clock frequencies f_{clk}, reduce device dimensions d and increase die area. Reduced device dimensions and increased die areas result in increased total switching capacitance C_{tot}. Therefore, as indicated in figure 3.28, the functional power dissipation for irreversible CMOS logic increases and reliability is endangered by increased electrical fields E. To keep power consumption and power density within limits, reducing the voltage swing U_{sw} is very effective as a result of the quadratic dependency. However, lower supply voltages increase propagation delays τ_{pd}, caused by lower saturation currents, affecting speed performance. To compensate for this nuisance threshold voltages and gate oxide thicknesses are reduced. Now the heart of the matter is reached, since weak-inversion currents increase exponentially with reduced threshold voltages. During standby periods weak-inversion currents become dominant, because functional and short-circuit power dissipation are non-existent. Especially for mobile, i.e. battery powered, applications increased standby power reduces battery lifetime. Therefore, the above stated problem will be called the speed performance standby power conflict.

Trends indicate that the functional as well as the parasitical power dissipation increase per generation, stressing the need for power reduction techniques. More and more desktop applications become mobile, and therefore dependent on batteries. On the one hand power reduction in general increases battery lifetime. On the other hand it saves the environment and money by reducing energy demands, which is not only valid for mobile but also for desktop applications. Techniques reducing the functional power dissipation will be presented in chapter 4. Chapter 5 discusses techniques to reduce the parasitical power dissipation.

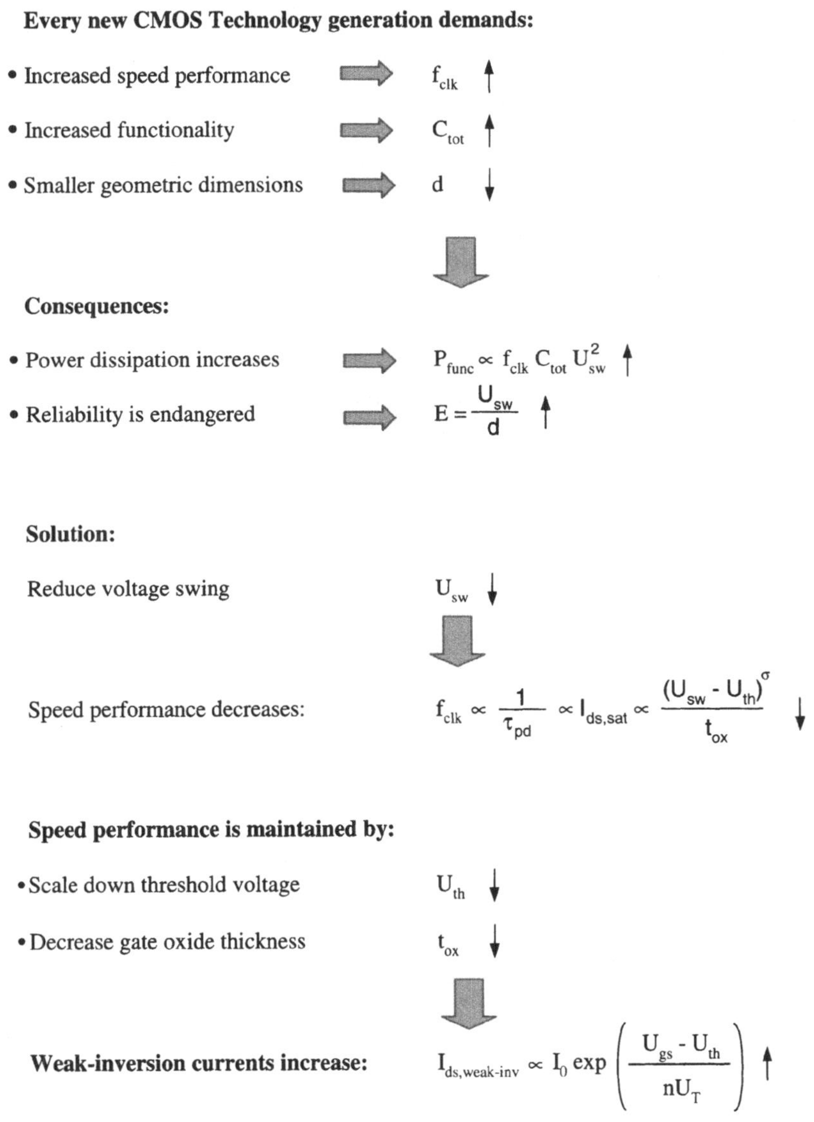

Every new CMOS Technology generation demands:

- Increased speed performance $\Longrightarrow$ f_{clk} ↑

- Increased functionality $\Longrightarrow$ C_{tot} ↑

- Smaller geometric dimensions $\Longrightarrow$ d ↓

Consequences:

- Power dissipation increases $\Longrightarrow$ $P_{func} \propto f_{clk}\, C_{tot}\, U_{sw}^2$ ↑

- Reliability is endangered $\Longrightarrow$ $E = \dfrac{U_{sw}}{d}$ ↑

Solution:

Reduce voltage swing U_{sw} ↓

Speed performance decreases: $f_{clk} \propto \dfrac{1}{\tau_{pd}} \propto I_{ds,sat} \propto \dfrac{(U_{sw} - U_{th})^\sigma}{t_{ox}}$ ↓

Speed performance is maintained by:

- Scale down threshold voltage U_{th} ↓

- Decrease gate oxide thickness t_{ox} ↓

Weak-inversion currents increase: $I_{ds,weak\text{-}inv} \propto I_0 \exp\left(\dfrac{U_{gs} - U_{th}}{nU_T} \right)$ ↑

Figure 3.28.   Trends in power dissipation for irreversible CMOS logic.

3.5   Conclusions

The power dissipation of a computational process depends on the physical process on which it is based. Two types of physical processes are distinguished:

- reversible processes;

- irreversible processes.

A reversible process distinguishes itself from an irreversible process by the ability to traverse all states in opposite direction in such a way that it dissipates no power and does not increase the entropy of the universe, when returning to the initial state.

Reversible logic returns to its initial state, after performing logical operations, via the reverse sequence of states. When these logical operations are physically being performed quasi-statically the process will be physically reversible. Practical reversible logic however, will have to perform logical operations in a limited amount of time, i.e. non quasi-statically. Although the power dissipation can be reduced substantially, they will dissipate power and therefore become physically irreversible.

Irreversible logic does not traverse the state sequences in the reverse order to return to the initial states after performing logical operations, and therefore it always is based on physical irreversible processes: logical irreversibility means physical irreversibility. For an irreversible process the theoretical minimal energy loss per logic bit equals $kT \ln 2$.

Adiabatic processes are not synonymous for lossless, because they are not necessarily reversible. However, physically reversible could be synonymous for lossless.

The majority of present-day computational systems consists of irreversible logic blocks, produced in CMOS technology. Hence, all power delivered to these systems will eventually be dissipated.

The total power dissipation of a digital CMOS circuit is divided into two classes:

- functional power;

- parasitical power.

The functional power is the power needed to just change the internal states, i.e. the state-capacitors' charges, of a digital CMOS circuit on behalf of information processing. The functional power dissipation equals:

$$\overline{P}_{func} = N\alpha_{eq}f_{clk,eq}E_{cycle,eq}$$

The parasitical power is power which is dissipated and dominant when the circuit is idle, defined as leakage power, or power which could be dissipated during state transitions without attributing to the actual changes of the internal states, defined as short-circuit power. The parasitical power dissipation is divided into two main groups, depending on whether the leakage is device or circuit related:

- leakage power dissipation;

- short-circuit power dissipation.

The leakage power dissipation is device related parasitical power, which is dissipated even when the circuit is idle.

Short-circuit power dissipation is caused by rail-to-rail currents during state transitions, without attributing to the actual changes of the internal states, and is only apparent in static CMOS logic circuits. In contrast to the leakage power it is circuit related parasitical power.

The leakage power is divided into three sub-groups:

- channel leakage current;

- diode leakage current;

- gate leakage current.

The channel leakage current consists of drain-source currents which are present even when gate-source voltages are zero. Three effects contributing to the channel leakage current are distinguished:

- weak-inversion current;

- drain-induced barrier lowering (DIBL) current;

- channel edge current.

Weak-inversion is the region of operation between depletion and moderate inversion, where the relation between the gate voltage and the drain-source current is exponential. The weak-inversion current, although applicable in low power analog applications, causes unwanted power dissipation in digital applications.

DIBL is the increase of the drain-source current caused by the increase in drain potential. Increasing the drain potential lowers the potential barrier between the source and the channel, making it easier for the source to inject carriers into the channel.

Channel edge currents in shallow trench isolated (STI) technologies are caused by the two dimensional influence of the gate at the edge of the channel, which lowers the effective threshold voltage at the edge compared to the center of the channel. Carefully designed trenches exhibit negligible edge currents compared to the weak-inversion currents of main channels.

The diode leakage current consists of source and drain pn-junction leakage currents. The diode leakage currents are divided into two components:

- reverse bias leakage;

- gate-induced drain leakage(GIDL).

The reverse bias leakage is the current through a reversed biased pn-diode, including the reverse saturation and generation current.

GIDL is a leakage current from drain to substrate, caused by high electric fields between gate and drain. The high electric field causes a depletion layer in the drain region in which tunneling assisted carrier pairs are created. These carriers are transported to the drain and the channel according to their polarity, thereby causing the GIDL leakage current.

The gate leakage current consist of direct tunneling current components and is caused by high electric fields across the gate oxide. The use of ultrathin gate oxides enhances device performance, but opens the door wide to gate leakage currents. Gate oxides of less than 2 nm will be required for sub-100 nm generation MOS transistors, causing gate oxide tunneling. Two gate tunneling current components can be distinguished:

- gate-to-channel direct tunneling current;

- source and drain extension-to-gate overlap tunneling current.

The gate-to-channel direct tunneling current is dominant below the channel flat-band voltage, in a n^+ polysilicon gate NMOS transistor.

The source and drain extension-to-gate overlap tunneling current is dominant between the channel flat-band voltage and the source and drain extension (SDE) flat-band voltage. In a n^+ polysilicon gate NMOS transistor the former flat-band voltage equals approximately -1 volt and the latter approximately zero. For large electric fields these tunneling currents dominate the drain leakage current for voltages between about 0 volt and -1 volt. Below that GIDL becomes dominant.

In order to reduce functional power dissipation and maintain device reliability while decreasing device dimensions, logic voltage swings need to be reduced. Since this degrades speed performance threshold voltages will have to be reduced as well, leading to increased weak-inversion currents. Weak-inversion currents become dominant during standby periods. The speed performance standby power conflict has become a fact.

4

REDUCTION OF FUNCTIONAL
POWER DISSIPATION

In chapter 3 reversible and irreversible processes as well as reversible and irreversible logic have been discussed. From this discussion it became clear that logical reversibility and quasi-statical changes of states are necessary boundary conditions for physical reversibility. All practical systems are physically irreversible, because they either consist of reversible logic and are non-quasi static or they consist of irreversible logic. Moreover, practical logic blocks are built with non-ideal devices exhibiting non-recoverable parasitical power dissipation. Only a part of the energy delivered to the reversible logic can be returned to the power supply afterwards, because the logical operations have to be performed in a limited amount of time. With irreversible logic all the energy delivered by the source is eventually converted into heat. One way or the other functional power, apart from parasitical power, will be dissipated. Hence, it is worthwhile finding opportunities to reduce it to a minimum.

In section 3.2 an expression for the functional power dissipation has been determined. According to this expression the functional power dissipation per circuit node equals:

$$\frac{\overline{P}_{func}}{N} = \overline{\alpha} f_{clk,eq} \overline{E}_{cycle} \tag{4.1}$$

The parameters in this expression will be addressed, in section 4.1 to 4.3, as a guide to discuss power reduction techniques reducing the functional power dissipation.

4.1 Node transition-cycle activity factor

The node transition-cycle activity factor, $\overline{\alpha}$, represents the average number of node transition-cycles in a clock period. Charging and discharging a circuit

53

node equals one transition cycle. The node transition-cycle activity factor will be divided into two parts:

- the functional part;

- the parasitical part.

The functional part $(0 < \overline{\alpha}_{func} \leq 1)$ depends on the algorithm, which means that algorithms should be optimized in such a way that the average node transition-cycle activity is minimal. Moreover, it depends on the logic function, i.e. AND, OR, or EXOR, the signal statistics and the choice of logic style, e.g. static, dynamic or pass-transistor logic [8].

The parasitical part consists of a number of extra transitions, i.e. glitching transitions, which are caused by the signal skews, i.e. due to different arrival times of the input signals, the signal reflections and the signal statistics [8]. The parasitical part can be reduced on the circuit level, among others by balancing the delays of the paths, or on the transistor level by gate sizing [39]. CMOS-gates exhibit low pass filtering for signals with a period shorter than the propagation delay of the gate. Hence, a glitch will not occur if the arrival times of input signals do not differ more than the propagation delay of the gate.

Synchronous circuits require special design effort and clock gated circuitry, i.e. flip-flops are added to fix the signal skew via the logic depth. Self-timed circuits [40] inherently avoid redundant transitions. For this reason, self timed circuits have attracted more attention in recent years, particularly in areas where the computational complexity is strongly data dependent [41].

4.2   Clock frequency

The clock frequency is one of the main parameters influencing the functional power dissipation. Power reduces proportional when lowering the clock speed (equation 4.1). However, reducing just the clock frequency also reduces the data processing speed and therefore the throughput. To maintain throughput for decreased clock frequencies, parallelisation and pipelining, discussed in section 4.3.3, can bring deliverance. The clock frequency can be controlled in combination with the supply voltage as a function of the work load. This will be addressed in section 4.3.3, where the supply voltage aspects are elaborated. Controlling the clock frequency as a function of temperature, can be applied to protect the chip against overheating. Of course the throughput is also influenced by this.

4.3   Transition-cycle energy

In a CMOS logic circuit state information is represented as charge on a circuit node, i.e. gate capacitor. Regarding the voltage swing U_{sw} of the circuit node with respect to the power supply voltage U_{dd}, two situations can be distinguished:

- $U_{sw} = U_{dd}$;

- $U_{sw} \neq U_{dd}$.

When the voltage swing equals the supply voltage, charging circuit node j with capacitance $C_{L,j}$ instantaneously with a voltage step $U_{dd,ij}$ in clock period i, the energy delivered by the supply equals:

$$
\begin{aligned}
E_{up,ij} &= \int_0^\infty U_{dd,ij}\, i_{C_L}(t)\, dt \\
&= \int_0^{U_{dd,ij}} U_{dd,ij}\, C_{L,j}\, d\, u_{C_L}(t) \\
&= C_{L,j} U_{dd,ij}^2
\end{aligned}
\tag{4.2}
$$

in which $i_{C_L}(t)$ and $u_{C_L}(t)$ are the current through and the voltage across the node capacitance, respectively. The supply voltage $U_{dd,ij}$ is fixed for node j during clock period i. Discharging the node capacitance instantaneously with a voltage step, i.e. short circuiting it, no energy will be restored in the supply. Therefore, the energy restored in the supply equals:

$$
E_{dn,ij} = 0
\tag{4.3}
$$

The total net dissipated energy during a complete charging and discharging cycle equals:

$$
E_{cycle,ij} = E_{up,ij} + E_{dn,ij} = C_{L,j} U_{dd,ij}^2
\tag{4.4}
$$

Charging and discharging instantaneously with voltage steps equal to U_{dd}, i.e. non-quasi statically, is common practice in static CMOS logic. However, techniques which approximate quasi-statical state changes are available and will be discussed in section 4.3.1. These techniques provide for energy restoration of the node charge to the supply after evaluation of the state of the circuit node. As a part of the energy delivered by the source can be restored afterwards, the total net transition-cycle energy dissipated in the circuit will be expressed as:

$$
E_{cycle,ij} = (1 - \eta_j) C_{L,j} U_{dd,ij}^2
\tag{4.5}
$$

in which η_j represents the reversibility factor of node j, which is the ratio between the energy restored by and delivered to the energy source. From equation 4.5 the average cycle energy $\overline{E}_{cycle}$ will be determined. The reversibility factor depends on the circuit properties and the form of the energy source signal, as will be explained in section 4.3.1. The supply voltage U_{dd} is an independent random variable. However, the parameters η and C_L are dependent, since both depend on circuit properties. Averaging the cycle energy over all circuit nodes and clock periods, as has been presented by equation 3.24 on page 22, the average cycle energy per circuit node can be expressed as:

$$\overline{E}_{cycle} \; = \; \frac{\sum\limits_{i=1}^{n} \sum\limits_{j=1}^{N} E_{cycle,ij}}{nN} \; = \; \frac{\sum\limits_{i=1}^{n} \sum\limits_{j=1}^{N} (1 - \eta_j)\, C_{L,j} U_{dd,ij}^2}{nN} \tag{4.6}$$

$$= \; \overline{(1 - \eta)\, C_L \, U_{dd}^2} \qquad\qquad 0 \leq \eta \leq 1$$

in which n represents the number of clock periods and N indicates the total number of circuit nodes. In this case the voltage swing of a circuit node, U_{sw}, equals the supply voltage.

A substantial amount of energy can be saved when the voltage swing of a circuit node is lower than the supply voltage. Reducing the voltage swing, two cases can be distinguished. The voltage swing is either derived from the supply voltage via a "lossy" or via an energy efficient power converter. In case of a lossy power conversion, as depicted in figure 4.1 [42], the transition-cycle energy becomes:

$$\overline{E}_{cycle} \; = \; \int_0^\infty (1 - \eta)\, U_{dd}\, i_{C_L}(t)\, dt \tag{4.7}$$

$$= \; U_{dd} \int_0^{U_{sw}} \overline{(1 - \eta)\, C_L}\; d\, u_{C_L}(t)$$

$$= \; \overline{(1 - \eta)\, C_L} U_{dd} \overline{U}_{sw} \qquad\qquad 0 \leq \eta \leq 1$$

It has been assumed here that the supply voltage U_{dd} is fixed.

The power conversion losses, within the energy supply source, can be diminished substantially by the use of energy efficient power converters, i.e. switching power supplies [43]. For a 100% efficient power supply the transition-cycle energy becomes:

$$\overline{E}_{cycle} = \overline{(1 - \eta)\, C_L \, U_{sw}^2} \qquad\qquad 0 \leq \eta \leq 1 \tag{4.8}$$

It should be noted that η only relates to the energy efficiency of the logic circuit.

For reversible logic switching quasi-statically, η equals 1, because the process is physically reversible and all energy will be recovered at the end of each logical operation, as has been discussed in section 3.1. However, η decreases for deviations from the quasi-static case, since only a part of the energy can be recovered. For irreversible logic η always equals 0, because no energy is recovered.

From the previous discussion it has become clear that only reversible logic creates an opportunity to let η approach 1. The reversibility factor and CMOS

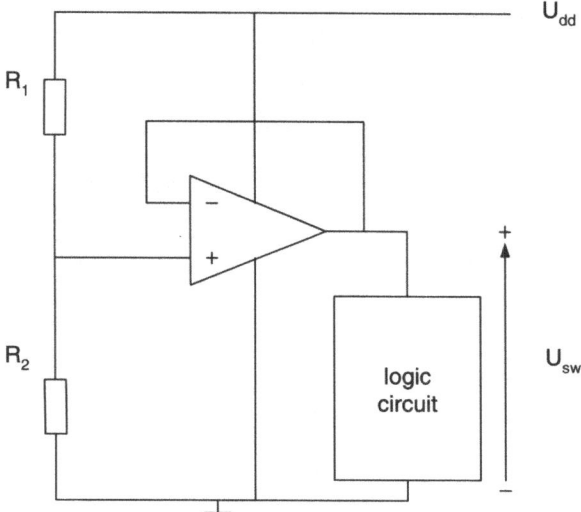

Figure 4.1. "Lossy" power converter used to reduce the voltage swing of a circuit node.

reversible logic will be discussed in section 4.3.1. Reduction of the load capacitance, discussed in section 4.3.2, and the reduction of the voltage swing, discussed in section 4.3.3, reduce the functional power dissipation for both reversible and irreversible logic.

4.3.1 Reversibility factor

In CMOS digital circuits the information contained by a circuit node is represented in the form of charge on a gate capacitance. Changing the logic state of a circuit node comes down to charging or discharging the gate capacitance. Changing the amount of charge quasi-statically, i.e. infinitely slowly, the process becomes physically reversible ($\eta = 1$) as has been discussed in section 3.1.1. However, logic operations have to be performed in a finite amount of time, i.e. non-quasi statically, causing loss of energy ($\eta < 1$).

The reversibility factor depends on the signal form of the energy source and the circuit properties, since the ratio between signal dynamics and circuit dynamics determine how close the quasi-statical case is approximated. To determine the relation between the energy loss during charge transport to and from the circuit node and the corresponding signal form of the energy source, the RC equivalent of the non-inverting output Q of the dual rail reversible AND gate of figure 4.2 [44] will be considered. Depending on the input signals A and B and their inverse counterparts $\overline{A}$ and $\overline{B}$ the energy source $u_{sw}(t)$ charges either the node capacitance belonging to Q or $\overline{Q}$. After the values of the outputs have been stabilized and evaluated the source $u_{sw}(t)$ discharges either the node capacitance belonging to Q or $\overline{Q}$; the charge on the relevant node capacitance

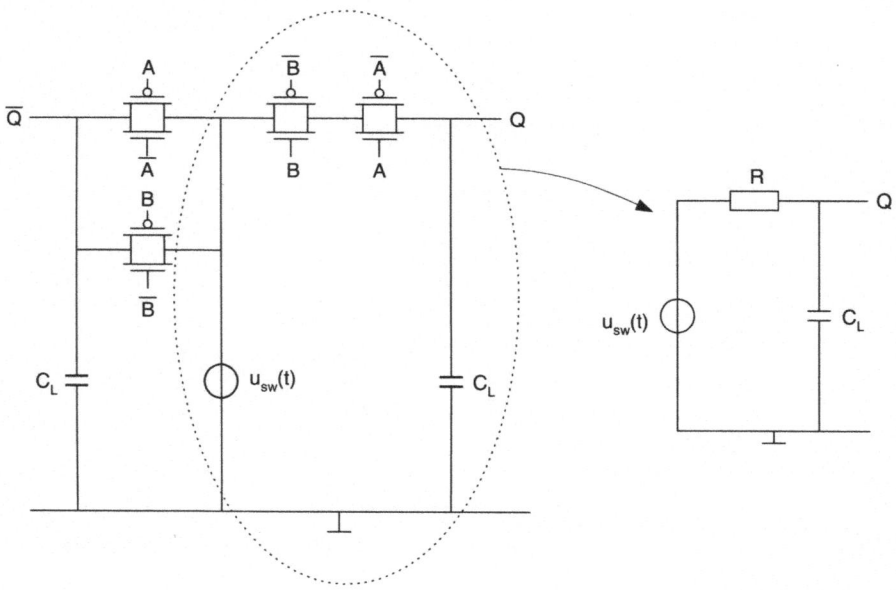

Figure 4.2.   A reversible AND gate and the RC circuit equivalent of the non-inverting output Q.

C_L is restored in the energy source. In a chain of logic gates the outputs of this AND are connected to the inputs of the next reversible logic gate. When the outputs of the AND are stable, the next reversible logic gate will be charged by a time-delayed energy source $u_{sw}(t - \tau_d)$. After the charging phase of the last reversible logic gate in the chain, its outputs will be stored in a latch. Figure 4.3 shows a chain of arbitrary reversible logic gates. Each chain of reversible logic gates is connected in between latches (not shown in figure 4.3), to keep the inputs, of the first logic gate in the chain, stable from the charging phase until the discharging phase, and to store the results of the evaluation of the last outputs in the chain. The evaluation phase is followed by the discharging phases of the logic gates by the energy sources in the reverse order; during the discharge phase of a logic gate its inputs have to remain stable.

In sections 4.3.1.1 to 4.3.1.3 three charging techniques are discussed to charge and discharge the circuit nodes by the energy supply source:

- ramp-wise charging;

- step-wise charging;

- resonant charging.

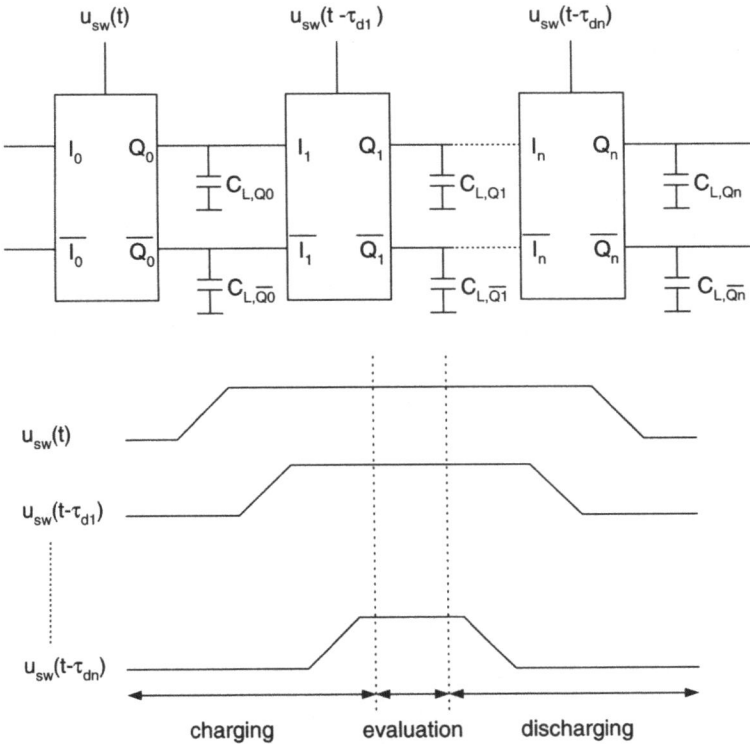

Figure 4.3. A chain of reversible logic gates.

4.3.1.1 Ramp-wise charging

The quasi-static state transitions can be best approximated by a ramp-shaped circuit node signal $u_{sw}(t)$, i.e. *ramp-wise charging* and discharging, see figure 4.4. The input energy source signal $u_{sw}(t)$, during the charging phase, equals:

$$u_{sw}(t) = \frac{U_{sw}}{\tau_r} t \epsilon(t) - \frac{U_{sw}}{\tau_r}(t - \tau_r) \epsilon(t - \tau_r) \qquad (4.9)$$

in which $\epsilon(t)$ represents the unit step function at $t = 0$, whereas $\epsilon(t - \tau_r)$ is a unit step at $t = \tau_r$, with τ_r the rise time and fall time of the ramp. For the voltage u_R across the resistor in the RC equivalent of figure 4.2 holds:

$$\begin{aligned} u_R(t) = & \frac{U_{sw}\,\tau}{\tau_r}\left[1 - \exp\left(-\frac{t}{\tau}\right)\right]\epsilon(t) - \\ & \frac{U_{sw}\,\tau}{\tau_r}\left[1 - \exp\left(-\frac{t - \tau_r}{\tau}\right)\right]\epsilon(t - \tau_r) \end{aligned} \qquad (4.10)$$

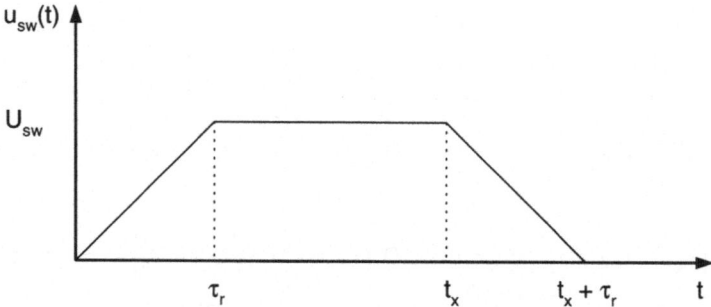

Figure 4.4.   Ramp-shaped energy source signal.

in which τ indicates the product of R and C_L. The energy dissipated in the resistor, E_R, equals:

$$E_R = \int_0^\infty \frac{(u_R(t))^2}{R} \, dt \qquad (4.11)$$

$$= \frac{(U_{sw}\,\tau)^2}{R\tau_r} \left[1 - \frac{\tau}{\tau_r} \left(1 - \exp\left(-\frac{\tau_r}{\tau}\right) \right) \right]$$

The energy dissipated with charging *and* discharging one circuit node during one clock period equals:

$$E_{cycle} = 2E_R \qquad (4.12)$$

The voltage swing U_{sw} is an independent random variable. Averaging the dissipated energy over all circuit nodes and clock periods, the cycle energy for the ramp-wise case can be represented as:

$$\overline{E}_{cycle} = 2\overline{E}_R = 2 \; \overline{\frac{\tau}{\tau_r} \left[1 - \frac{\tau}{\tau_r} \left(1 - \exp\left(-\frac{\tau_r}{\tau}\right) \right) \right] C_L \; \overline{U_{sw}^2}} \qquad (4.13)$$

Comparison of equation 4.13 and the general equation 4.8 yields the reversibility factor for the ramp-wise case:

$$\eta = 1 - \frac{2\tau}{\tau_r} \left[1 - \frac{\tau}{\tau_r} \left(1 - \exp\left(-\frac{\tau_r}{\tau}\right) \right) \right] \qquad (4.14)$$

Ramping up and down, i.e. a complete transition cycle, a circuit node quasi-statically yields a reversibility factor equal to:

$$\lim_{\tau_r \to \infty} \eta = 1 \qquad (4.15)$$

Applying an infinitesimal small rise time in the ramp-wise case results in a unit step. The reversibility factor follows from equation 4.14, by substitution of the Taylor series expansion for $\exp\left(-\frac{\tau_r}{\tau}\right)$:

$$\lim_{\tau_r \to 0} \eta = \lim_{\tau_r \to 0} 1 - \frac{2\tau}{\tau_r}\left[1 - \frac{\tau}{\tau_r}\left(\frac{\tau_r}{\tau} - \frac{\tau_r^2}{2\tau^2} + \cdots + \frac{\tau_r^n}{n!\tau^n}\right)\right] \qquad (4.16)$$

$$= \lim_{\tau_r \to 0} 1 - 2\tau\left[\frac{1}{\tau_r} - \frac{\tau}{\tau_r^2}\left(\frac{\tau_r}{\tau} - \frac{\tau_r^2}{2\tau^2} + \cdots + \frac{\tau_r^n}{n!\tau^n}\right)\right] = 0$$

The unit step charging and discharging of circuit nodes is common place in irreversible logic circuits, like for instance static CMOS circuits. Therefore, $\eta = 0$ and the amount of energy dissipated in a complete cycle equals $\overline{C}_L \overline{U}_{sw}^2$.

Figure 4.5 shows the reversibility factor of equation 4.14 as a function of $\frac{\tau_r}{\tau}$. From this plot it becomes clear that for higher clock rates, i.e. smaller rise and fall time values τ_r, the efficiency, and thus the reversibility factor, decreases rapidly.

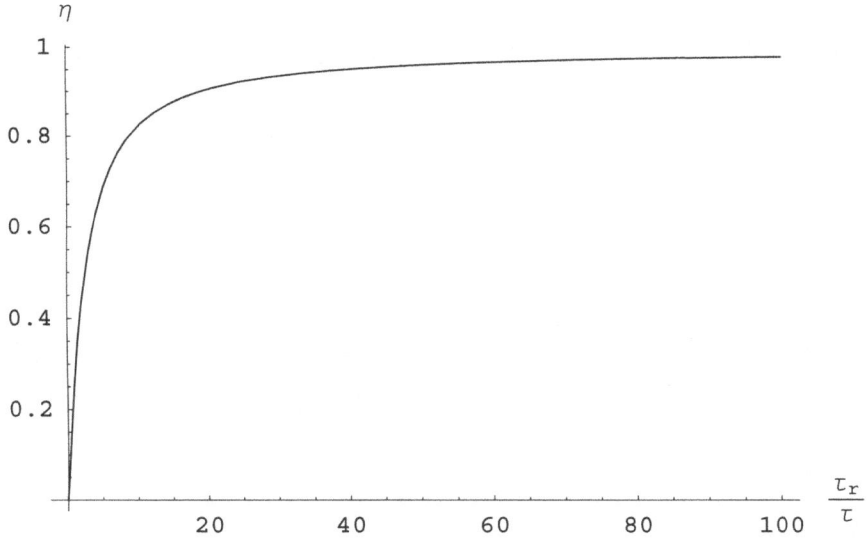

Figure 4.5. The reversibility factor for a ramp-shaped energy source signal as function of $\frac{\tau_r}{\tau}$.

4.3.1.2 Step-wise charging

The ramp can be divided into p equal steps, resulting in the *step-wise charging* technique [45], which approximates the ramp-wise charging technique. In this technique the supply line is charged and discharged in p steps of $\frac{U_{sw}}{p}$ volt each. The energy delivered by the power supply source E_{up} when charging a load capacitor C_L equals:

$$E_{up} = \int_0^{\frac{U_{sw}}{p}} C_L \frac{U_{sw}}{p} \, dU_{C_L} + \tag{4.17}$$

$$+ \int_{\frac{U_{sw}}{p}}^{2\frac{U_{sw}}{p}} 2C_L \frac{U_{sw}}{p} \, dU_{C_L} + \cdots + \int_{(p-1)\frac{U_{sw}}{p}}^{U_{sw}} C_L U_{sw} \, dU_{C_L}$$

$$= \left(\frac{p+1}{p}\right) \frac{1}{2} C_L U_{sw}^2$$

The energy dissipated during the charging process equals:

$$E_{R,up} = E_{up} - E_{C_L} \tag{4.18}$$

$$= \left(\frac{p+1}{p}\right) \frac{1}{2} C_L U_{sw}^2 - \frac{1}{2} C_L U_{sw}^2$$

$$= \frac{C_L U_{sw}^2}{2p} \tag{4.19}$$

in which E_{C_L} is the energy stored in the load capacitor. The energy restored in the power supply source E_{dn} when discharging capacitor C_L equals:

$$E_{dn} = \int_{U_{sw}}^{\frac{(p-1)U_{sw}}{p}} (p-1)C_L \frac{U_{sw}}{p} \, dU_{C_L} + \cdots + \tag{4.20}$$

$$+ \int_{3\frac{U_{sw}}{p}}^{2\frac{U_{sw}}{p}} 2C_L \frac{U_{sw}}{p} \, dU_{C_L} + \int_{2\frac{U_{sw}}{p}}^{\frac{U_{sw}}{p}} C_L \frac{U_{sw}}{p} \, dU_{C_L}$$

$$= \left(\frac{p-1}{p}\right) \frac{1}{2} C_L U_{sw}^2$$

The energy dissipated during the discharging process equals:

$$E_{R,dn} = E_{C_L} - E_{dn} \tag{4.21}$$

$$= \frac{1}{2} C_L U_{sw}^2 - \left(\frac{p-1}{p}\right) \frac{1}{2} C_L U_{sw}^2$$

$$= \frac{C_L U_{sw}^2}{2p} \tag{4.22}$$

Summation of equation 4.18 and 4.21 yields the cycle energy:

$$\overline{E}_{cycle} = \overline{E_{R,up} + E_{R,dn}} \tag{4.23}$$

$$= \frac{\overline{C_L U_{sw}^2}}{p}$$

In this case η equals $1 - 1/p$.

4.3.1.3 Resonant charging

The *resonant charging* technique uses an inductance, L, in the energy supply source, which form an RLC resonator with the RC load of the logic circuit [46, 47, 48, 49]. In literature many implementations for the resonating supply source can be found. The circuit of figure 4.6 [46] will be used to explain the resonant charging technique. This supply source consists of an inductance L and a tank capacitor, C_T. The signals, SW_1 and SW_2 control the charging and discharging sequence of the circuit node. Although the ramp-wise

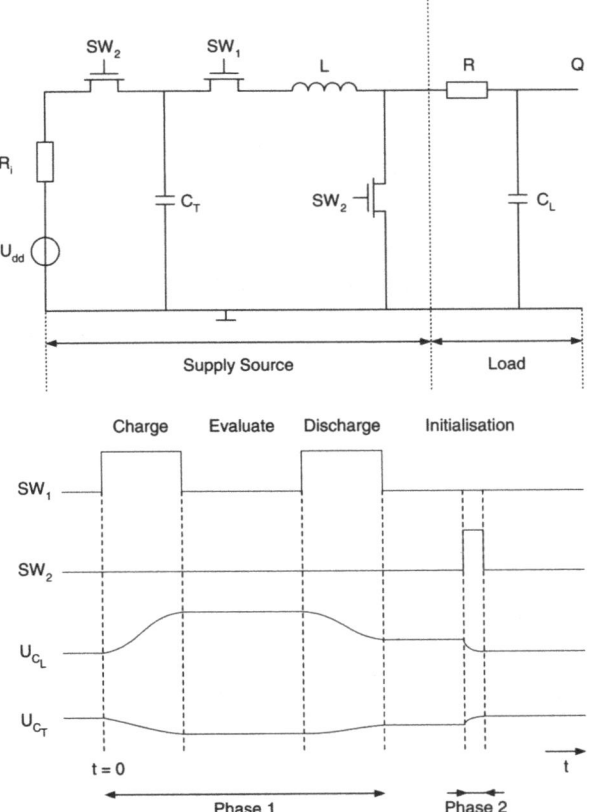

Figure 4.6. Resonant charging technique [46].

charging technique approximates the quasi-static state transitions best, an oscillating wave form can be generated more easily in practice. A complete cycle consists of two phases. *The first phase* comprises charging, evaluation and discharging of the circuit node capacitance C_L. During this phase the switch controlled by SW_1 is conducting, whereas the switches controlled by SW_2 are non-conducting, see figure 4.6. Switches are controlled only when currents are zero. During the evaluation period switch SW_1 is isolated to stabilize the output voltage by preventing charge from flowing to or from the output node C_L. During evaluation the state of the output node C_L is sampled by a latch, not shown in figure 4.6. The first phase starts at $t = 0$ with an initial energy stored in the tank capacitance:

$$E_i = \frac{1}{2} C_T U_{dd}^2 \tag{4.24}$$

After state evaluation and discharging the load capacitance the dissipated energy equals:

$$E_{cycle,1} \overset{\Delta}{=} (1 - \eta_1) E_i \tag{4.25}$$

in which η_1 represents the reversibility factor of the first phase. In *the second phase* the small rest energy on C_L is removed and the energy lost during the first phase is restored in the tank capacitance. During this phase the switch controlled by SW_1 is non-conducting, whereas the switches controlled by SW_2 are conducting. In this case the energy of the tank capacitance is changed with the step-wise charging technique from $\eta_1 E_i$ to E_i. From this action the potential of the tank capacitance changes from $\sqrt{\eta_1} U_{dd}$ to U_{dd}. Consequently, the power supply source U_{dd} has to deliver an energy E_{source} equal to:

$$
\begin{aligned}
E_{source} &= \int_{\sqrt{\eta_1} U_{dd}}^{U_{dd}} U_{dd} C_T \; d\, U_{C_T} \\
&= (1 - \sqrt{\eta_1})\, C_T U_{dd}^2
\end{aligned}
\tag{4.26}
$$

Since the amount added to the energy stored in the tank capacitance equals $E_{cycle,1}$, the energy dissipated in the internal resistance R_i of the voltage source U_{dd} in the second phase $E_{cycle,2}$ equals:

$$
\begin{aligned}
\overline{E}_{cycle,2} &= \overline{E_{source} - E_{cycle,1}} \\
&= \overline{(1 - \sqrt{\eta_1})^2 \, E_i}
\end{aligned}
\tag{4.27}
$$

The total cycle energy equals the sum of the cycle energies of both phases:

$$
\begin{aligned}
\overline{E}_{cycle} &= \overline{E}_{cycle,1} + \overline{E}_{cycle,2} \\
&= \overline{(1 - \sqrt{\eta_1})\, C_T\, U_{dd}^2}
\end{aligned}
\tag{4.28}
$$

To determine the reversibility factor η_1 for the resonant charging technique, the initial energy, i.e. E_i, and final energy, i.e. $\eta_1 E_i$, of the tank capacitance will be regarded. For this purpose the voltage across the tank capacitance as a function of time is derived:

$$u_{C_T}(t) = \frac{1}{C_T} \int i(t) \; dt \tag{4.29}$$

The current through the series connection of C_T, L, R and C_L is represented by $i(t)$, which can be calculated with the second-order differential equation of the circuit of figure 4.6, and equals:

$$i(t) = \frac{U_{dd}}{\lambda L} \exp(-\zeta t) \sin(\lambda t) \tag{4.30}$$

in which ζ and λ equal:

$$\zeta = \frac{R}{2L} \tag{4.31}$$

$$\lambda^2 = \omega_0^2 - \zeta^2$$

For the oscillation frequency ω_0 of the RLC resonator holds:

$$\omega_0 = \frac{1}{\sqrt{LC_s}} \tag{4.32}$$

in which C_s is the equivalent series capacitance of C_T and C_L. A necessary boundary condition for the circuit to perform an oscillation is:

$$0 < \zeta < \omega_0 \tag{4.33}$$

Substitution of equation 4.30 in equation 4.29 yields:

$$u_{C_T}(t) = \frac{C_T}{C_T + C_L} \left[1 + \frac{C_L \omega_0}{C_T \lambda} \exp(-\zeta t) \sin(\lambda t + \theta) \right] U_{dd} \tag{4.34}$$

The phase angle θ is represented by:

$$\theta = \arctan(\frac{\lambda}{\zeta}) \tag{4.35}$$

The voltages and thus the energies of the tank and load capacitances reach their extremes when the current $i(t)$ is zero. From equation 4.30 it can be seen that $i(t)$ is zero for $\lambda t = 0 + k\pi$. For even values of k the tank energy is maximal. When $t = 0$ the tank energy equals the initial energy $e_{C_T}(0) = E_i$, whereas for $t = \frac{2\pi}{\lambda}$ the tank energy reaches a maximum for the second time, just when the discharge period has been completed and most of the energy in C_L has been restored in C_T. Now the cycle energy of the first phase can be determined as the difference between these two energies:

$$E_{cycle,1} = e_{C_T}(0) - e_{C_T}\left(\frac{2\pi}{\lambda}\right) = \frac{1}{2} C_T U_{dd}^2 - \frac{1}{2} C_T u_{C_T}^2 \left(\frac{2\pi}{\lambda}\right) \tag{4.36}$$

For the tank voltage at $t = \frac{2\pi}{\lambda}$ holds:

$$u_{C_T}\left(\frac{2\pi}{\lambda}\right) = \frac{C_T}{C_T + C_L} \left[1 + \frac{C_L}{C_T} \exp\left(-2\pi \frac{\zeta}{\lambda}\right) \right] U_{dd} \tag{4.37}$$

Substitution of equation 4.37 in equation 4.36 yields:

$$E_{cycle,1} = \left[1 - \left(\frac{C_T + C_L \exp\left(-2\pi\frac{\zeta}{\lambda}\right)}{C_T + C_L}\right)^2\right] E_i \qquad (4.38)$$

The reversibility factor η_1 can be found by comparing equation 4.25 with equation 4.38 and equals:

$$\eta_1 = \left(\frac{C_T + C_L \exp\left(-2\pi\frac{\zeta}{\lambda}\right)}{C_T + C_L}\right)^2 \qquad (4.39)$$

The factor $\frac{\zeta}{\lambda}$ can be written in a form containing the quality factor Q_f of the RLC resonator:

$$\frac{\zeta}{\lambda} = \frac{1}{\sqrt{4Q_f^2 - 1}} \qquad (4.40)$$

The quality factor equals:

$$Q_f = \frac{\omega_0 L}{R} \qquad (4.41)$$

Substituting equation 4.40 in equation 4.39 yields for large values of Q_f:

$$\eta_1 = \left(\frac{C_T + C_L \exp\left(-\frac{\pi}{Q_f}\right)}{C_T + C_L}\right)^2 \qquad (4.42)$$

From this equation it can be seen that the reversibility factor increases for larger C_T and Q_f. Substitution of equation 4.42 in equation 4.28 yields the expression for the total average cycle energy, expressed in U_{dd}:

$$\overline{E}_{cycle} = \overline{\left[1 - \left(\frac{C_T + C_L \exp\left(-\frac{\pi}{Q_f}\right)}{C_T + C_L}\right)\right] C_T \, \overline{U_{dd}^2}} \qquad (4.43)$$

To be able to compare the cycle energy of the resonant charging technique with the ramp-wise and step-wise charging technique, equation 4.43 will be expressed in terms of the voltage swing U_{sw} and the load capacitance C_L. To accomplish this the relation between U_{dd} and U_{sw} is determined. As was the case with the voltage across the tank capacitance the voltage on the load capacitance reaches its extremes when the current $i(t)$ equals zero. The voltage across C_L equals:

$$u_{C_L}(t) = \frac{C_T}{C_T + C_L}\left[1 - \frac{\omega_0}{\lambda}\exp(-\zeta t)\sin(\lambda t + \theta)\right] U_{dd} \qquad (4.44)$$

The voltage swing U_{sw} is the voltage across C_L at the first maximum of $u_{C_L}(t)$, which is reached at $t = \frac{\pi}{\lambda}$. Therefore, U_{sw} can, for large Q_f, be expressed as:

$$U_{sw} = u_{C_L}\left(\frac{\pi}{\lambda}\right) = \frac{C_T}{C_T + C_L}\left[1 + \exp(-\frac{\pi}{2Q_f})\right]U_{dd} \qquad (4.45)$$

Substitution of equation4.45 in equation 4.43 yields:

$$\overline{E}_{cycle} = \overline{\frac{C_T + C_L}{C_s} \frac{(1 - \sqrt{\eta_1})}{\left(1 + \exp\left(-\frac{\pi}{2Q_f}\right)\right)^2} C_L U_{sw}^2} \equiv \overline{(1 - \eta)C_L U_{sw}^2} \quad (4.46)$$

in which η represents the overall reversibility factor. Note that the voltage swing depends on circuit parameters and therefore is not an independent random variable. Figure 4.7 indicates the relation between the overall reversibility factor and the quality factor, for different values of $\frac{C_T}{C_L}$. Negative values for η indicate

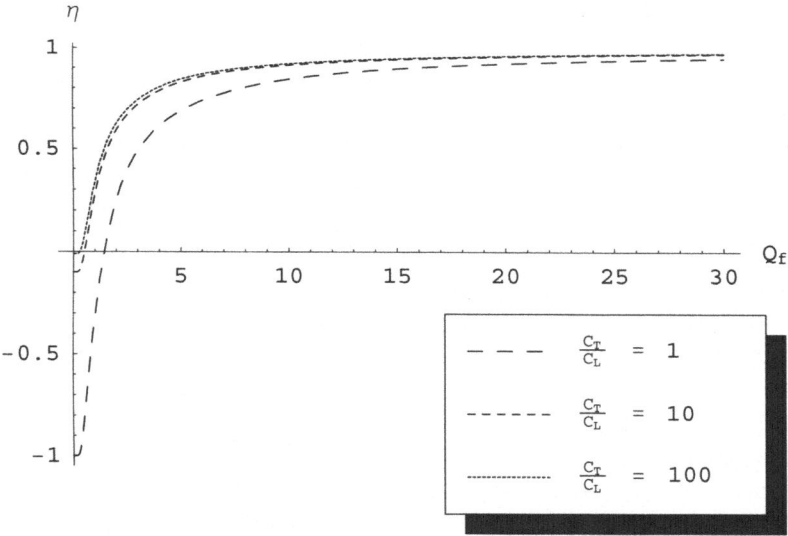

Figure 4.7. Overall reversibility versus quality factor for different values of $\frac{C_T}{C_L}$.

that an amount of energy larger than $C_L U_{sw}$ is dissipated during a complete cycle, which is a quite unfavorable setting since more energy is dissipated as in the case of an irreversible process. Increasing the quality factor increases the reversibility factor, hence decreasing the power dissipation.

From figure 4.7 it can be seen that for small quality factors small changes in this factor result in large changes of reversibility factors. For every new CMOS technology the total load capacitance, i.e. C_L, and the clock frequency, i.e. ω_0, increase. This leads to rapid decreasing reversibility factors, threatening future application of this technique. This can be explained as follows. For constant $\frac{C_T}{C_L}$, C_s increases with the same factor as C_L. From equation 4.32 it can be seen that L has to decrease for increasing C_s and ω_0. It is assumed that the resistance of the switches in the logic, i.e. R, are dominant. For constant R, a decreasing L reduces the quality factor Q_f (equation 4.41) of the RLC resonator, resulting in rapid declining reversibility factors for small quality factors.

4.3.2   Load capacitance

The total physical node capacitance, i.e. the load of a logic CMOS gate, will be divided into two parts:

- the transistor capacitance;

- the wiring capacitance.

These can be decreased by proper circuit and layout design, and technology scaling. Scaling of the node capacitance increases transistor density, and decreases the power dissipation and the circuit delays.

The transistor capacitance is formed by gate and junction capacitances. The gate capacitance, which is functional, consists of three components: the capacitance between the gate and the source, the gate and the drain, and the gate and the bulk region. The gate capacitance, C_g, can be approximated by:

$$C_g = C_{ox}' A_g = \frac{K_{ox}}{t_{ox}} W_{ch} L_{ch} \qquad (4.47)$$

Decreasing the gate capacitance not only gives an opportunity to increase transistor density, it also decreases the power dissipation per transistor. In the case of constant-field scaling W_{ch}, L_{ch}, and t_{ox} are scaled with a factor of $1/\kappa$, where $1/\kappa$ is about 0.7, i.e. 30% scaling [50, 51]. Consequently, C_{ox}' scales with κ, and C_g scales with $1/\kappa$. Decreasing the oxide thickness is necessary for the gate to keep enough grip on the channel, since for shorter channels the influence of the source and drain on the channel region increases. Gate oxide tunneling becomes a huge problem for oxide thicknesses less than 1 nm. New gate-insulator materials with higher dielectric permittivity, K_{ox}, would solve this discrepancy.

The junction capacitance is parasitical and consists of the reverse-biased pn-junction capacitance between the drain (or source) and the substrate. The junction capacitance, C_j, can be expressed as:

$$C_j = C_j' A_j = \frac{K_{Si}}{t_{dep}} \left(W_{ch} L_s + 2 W_{ch} t_j + 2 L_s t_j \right) \qquad (4.48)$$

in which t_j, t_{dep} and L_s are the junction depth, the thickness of the junction depletion layer and the length of the source or drain junction, respectively.

K_{Si} is the permittivity of silicon. Under constant-field scaling conditions the junction capacitance scales with a factor of $1/\kappa^1$ when the depletion layer thickness scales with $1/\kappa$, and the source and drain diffusion areas scale with $(1/\kappa)^2$. The depletion layer thickness is proportional to the square root of the ratio between the source or drain junction voltage and the substrate doping concentration. Since the source and drain junction voltages scale with $1/\kappa$ and the substrate doping concentration scales with κ, the depletion layer thickness therefore scales with $1/\kappa$.

The wiring capacitance is a parasitical component. The wire-to-wire capacitance within a metal layer is dominant compared to the wire-to-wire capacitance between layers, since the ratio between wire thickness t_w and wire width increases, wire pitches decrease and the ratio between the wire width and the inter-level dielectric thickness increases. The wire-to-wire capacitance within a metal layer can therefore be expressed as:

$$C_w = C_w' A_w = \frac{K_{ID}}{s_w} t_w L_w \qquad (4.49)$$

in which K_{ID} is the permittivity of the isolation dielectric material separating wires and s_w, t_w and L_w represent the wire spacing, thickness and length, respectively. The wiring capacitance can be minimized by scaling dimensions and exploiting (new) material properties, like for instance isolation material with a low relative permittivity. The wire spacing and the wire width scale with $1/\kappa$ for every technology generation. The wire thickness however, scales with approximately $1/\sqrt{\kappa}$, and the average wire length is predicted to scale with $1/\sqrt{\kappa}$ [51] as well. To extend the wiring scaling trend, i.e. $1/\kappa$, into the future, new inter-level dielectric materials need to be found of which the permittivity also scales with $1/\kappa$. This will become quite a challenge. Applying for instance fluorinated SiO_2 reduces the permittivity from $K_{ID} = 3.9K_0$ to $K_{ID} = 3.6K_0$ [51], where K_0 is the permittivity of free space. In this case the scaling factor of the dominant wire capacitance is much less than $1/\kappa$. For this reason the wire capacitance will not scale as aggressively as the transistors.

4.3.3 Voltage swing

As power consumption in CMOS circuits is proportional with the square of the voltage swing of a circuit node, it is clear that reduction of this parameter has a significant impact. Two voltage scaling techniques will be distinguished:

- static voltage scaling;

- dynamic voltage scaling.

---

[1]The relative benefit of Silicon On Insulator (SOI) CMOS technology will reduce, because the impact of junction capacitance diminishes with scaling [52].

Static voltage scaling implies the choice of an optimal fixed voltage swing, keeping technology, performance, and reliability in mind. Static voltage scaling will be discussed in section 4.3.3.1. *Dynamic voltage scaling* is a technique that varies the amplitude of the voltage swing as a function of the system parameters. Dynamic voltage scaling will be discussed in section 4.3.3.2.

When decreasing the voltage swing a limit will be reached eventually. What determines the lower limit to the voltage swing? Two limits can be determined. The first one is a theoretical limit, i.e. the absolute lowest voltage swing. The second one is a practical limit, i.e. determined by a reasonable speed of operation.

The theoretical lower limit on the voltage swing is determined with the minimum cycle energy and the minimum amount of charge. According to [53] and the discussion in section 3.1.1 the minimum irreversible transition-cycle energy equals:

$$E_{cycle,min} = 2E_{transition,min} = 2kT\ln 2 \qquad (4.50)$$

Therefore, the minimum voltage swing of a circuit node $U_{sw,min}$ becomes:

$$U_{sw,min} = \frac{E_{cycle,min}}{Q_{gate,min}} = \frac{E_{cycle,min}}{q} = 2\frac{kT}{q}\ln 2 \qquad (4.51)$$

The minimum possible gate charge $Q_{gate,min}$ is assumed to be the charge of a single electron, i.e. q. It can be concluded that the same minimum energy result is found for the single-electron case, as in the case of the one-atom gas discussed in section 3.1.1.

The practical lower limit on the voltage swing depends on the margin between the voltage swing and the device thresholds, which determines the speed of operation. As voltage swings approach device thresholds, gate propagation delays, τ_{pd}, increase rapidly. Equation 4.52 indicates a first order approximation of an inverter's propagation delay.

$$\tau_{pd} = R_{ds}C_L = C_L\frac{U_{sw}}{I_{ds,sat}} = C_L\frac{U_{sw}}{\frac{\beta}{2}(U_{sw} - U_{th})^\sigma} \qquad (4.52)$$

in which R_{ds} represents the impedance of the channel of an MOS transistor, C_L is the load capacitance of the inverter, and $I_{ds,sat}$ the drain-source saturation current. Correspondingly, the power savings increase to a point where the added overhead dominates any gains in power reduction from further voltage reduction, leading to the existence of an optimal voltage from an architectural point of view.

4.3.3.1   Static voltage scaling

Reduction of the signal voltage swing will decrease the output current of the MOS transistor, causing an increase in delay. Minimum power dissipation for constant throughput in a circuit is reached when paths become time critical. Non-time-critical paths can be made critical by lowering their signal voltage swing; multiple fixed voltages can be used within one circuit block [54]. Parallelisation enables further decrease of power dissipation while maintaining throughput. Two forms of parallelisation can be distinguished:

- parallel operations performed on the same signal;

- parallel operations performed on different signals, generated successively in time.

Parallel operations on the same signal, in literature sometimes referred to as hardware parallelisation, can be applied when an algorithm exhibits concurrency. Different operations on the same signal, which can be performed concurrently, are divided over parallelized hardware. Hence, the calculation times decrease, thereby increasing the margins of the critical paths. These margins can be exchanged for lower voltage swings and clock frequencies. In the case of general purpose hardware, N times parallelisation often means at least N times more hardware plus overhead; scaling down the clock frequency compensates for the increase in load capacitance, caused by the extra hardware. Hence, area is traded for lower power consumption through hardware parallelisation, while maintaining constant throughput.

Parallel operations performed on different signals, generated successively in time, in literature referred to as pipelining. Pipelining reduces the logic depth and thus the critical path. In case the logic depth can be reduced, pipelining may be applied. By the addition of extra registers in the critical path (figure 4.8a) the logic between these registers could operate slower while maintaining functional throughput. This implies that the voltage swing can be scaled to

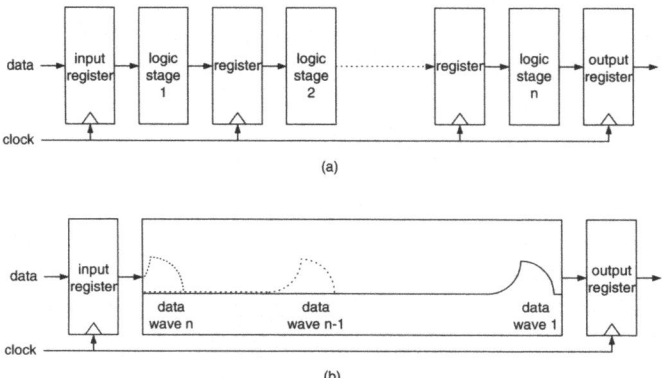

Figure 4.8. (a) Conventional pipeline, (b) wave pipeline.

slow down the circuits between the registers [54]. Notice that the latency is increased for the pipelined implementation. However, the area overhead for

the pipelined implementation is much smaller compared to hardware parallelisation, since only registers have to be added. A special pipelining technique
is wave pipelining [55]. In this pipeline the registers are removed and consec-
utive data elements move in separate successive waves through the different
sub-circuits (figure 4.8b). Extreme care has to be taken with respect to timing
aspects of each sub-circuit in the line. Combination of hardware parallelisation
and pipelining is an obvious extension [8].

The above-mentioned techniques can also be applied on systems with multiple voltage swings. In these systems different parts of a system operate on
different static voltages levels, depending on the speed requirements [6, 7].

4.3.3.2   Dynamic voltage scaling

In subsection 4.3.3.1 techniques were presented to scale down the voltage
swing of circuit nodes. However, the voltage swing itself remained static. It
is also possible to adapt the voltage swing dynamically, in order to process
data samples "just in time", implying minimum power consumption. Dynamic
voltage scaling is a power reduction technique, on the circuit level, that varies
the voltage swing, and can be subdivided into:

- parameter-controlled voltage scaling;

- data-controlled voltage scaling.

In a *parameter-controlled voltage scaling* system the information to control the
voltage swing can be delivered by internal and external parameters. Examples
of internal parameters are process variations and ageing effects. Examples of
external parameters are temperature, stress and radiation. Those parameters
influence the critical path of a circuit. The critical path can be imitated by
a ring oscillator with a delay versus voltage-swing behavior, identical to that
of the actual critical path [9, 56, 57]. This means that internal and external
parameter variations will influence delays in critical paths in the circuit and
in the ring oscillator identically. Changing the voltage swing of a ring oscillator changes the frequency; the ring oscillator becomes a Voltage Controlled
Oscillator (VCO). Hence, a possible implementation is to combine a regulated
supply and a loop filter in a Phase Lock Loop (PLL), as indicated in figure
4.9, to provide for a regulated voltage swing. The regulated supply powers the
ring oscillator, i.e. the VCO. The phase difference between the output signal
of the VCO and a reference clock frequency, e.g. the chip's general clock frequency, is used as a control signal for the regulated supply. Thus the voltage
swing is regulated in order to keep the clock frequency equal to the reference
frequency. A signal provided at the offset terminal of the divider creates the
opportunity to obtain a security margin or to control the voltage swing and
the clock frequency as a function of an external parameter, like for instance

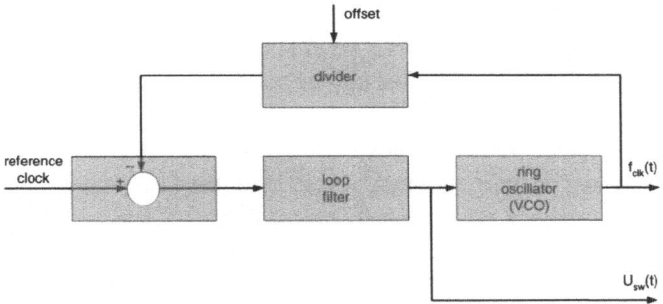

Figure 4.9. Regulation of the voltage swing with a Phase Lock Loop to make the clock frequency independent of internal and external parameter variations. The offset input enables both control of the voltage swing and the clock frequency.

temperature. This self-regulating circuit adjusts the internal voltage swing to the lowest value compatible with chip speed requirements, taking internal and external parameters into account.

In a *data-controlled voltage scaling* system extra control information is delivered by the workload, which influences the voltage swing $U_{sw}(t)$ via the offset control terminal, as depicted in figure 4.10. Examples of systems with workload-controlled voltage swings or "just-in-time processing systems" are synchronous systems containing adaptive power supplies, and asynchronous self-timed systems.

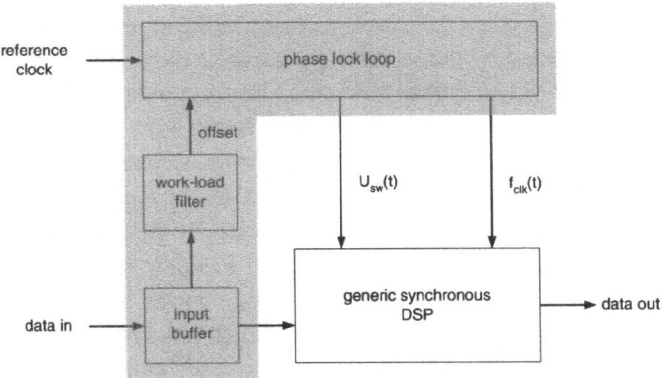

Figure 4.10. Voltage swing and clock frequency which are both adaptive to workload, internal and external parameters.

In the synchronous system set-up both the voltage swing and the clock frequency are workload dependent. Internal and external parameter variations can be compensated for as well. The control loop consists of a combination of a parameter-controlled and data-controlled voltage scaling system [10] (figure 4.10). The forward transfer from the data input via the input buffer and the workload filter to the regulated supply and the ring oscillator, within the PLL, compensates for the workload variations. The feed-back path from the ring oscillator to the phase detector compensates for temperature and process variations (figure 4.9).

Self-timed circuits (delay-insensitive circuits) [40, 41] are asynchronous, and the sequencing of their computations is determined by the data flow rather than by clock signals or other global control signals. The arrival of data at a successor latch triggers the predecessor latch, via an acknowledge signal, to let its data pass, as shown schematically in the pipeline of figure 4.11. In most

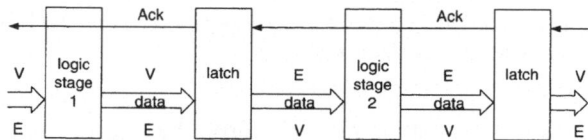

Figure 4.11. Delay-insensitive pipeline.

practical designs data is represented in a dual-rail code, i.e. the state of a data bit x is presented by two wires: x_t and x_f. The state is called empty (E) if both wires posses a low value. The state is called valid (V) when bit x is True, i.e. $x_t = 1$ and $x_f = 0$, or False, i.e. $x_t = 0$ and $x_f = 1$. A valid state is always preceded by an empty state and vice versa. A combination of both states is called a token. The token ripples through the pipeline from left to right, whereas the acknowledge signals ripple from right to left. A latch loads and holds a valid state when its successor latch in the pipeline holds the empty state, which is indicated by the incoming acknowledge being low. Similarly, a latch may load and hold an empty state when its successor latch holds the valid state, which is indicated by the acknowledge being low. The logic stage, i.e. the logic function block, performs operations on valid states only, empty states are passed unchanged. In the self-timed system set-up the voltage-swing level is, like in the synchronous system case, controlled by the amount of data in an input buffer. More data in the buffer results in a higher voltage swing, which speeds up the system. The input buffer also filters peaks in the workload. The set-up of figure 4.9 could also be used for self-timed systems, the loop with the ring oscillator is then used for compensation of temperature and process variations only. No system clock is needed, since the system is asynchronous. Consequently, variations in fabrication processes and operating conditions can be compensated for. The performance of the system depends on actual circuit delays, rather than on worst-case delays. A delay insensitive pipeline seems similar to the wave pipeline, except for the handshaking, which guaranties the integrity of the data under all propagation conditions.

Algorithms and applications that are particularly suited for dynamic voltage scaling are systems in which data comes in bursts like: sampled audio systems, floating-point units, general purpose CPU's, MPEG video (de)coders, speech recognizing and error correcting systems.

4.4 Conclusions

In all practical systems logical operations have to be performed in a limited amount of time, i.e. non-quasi statically. Therefore, the energy delivered to the reversible logic can be returned only partly to the power supply, after a logical operation has been performed. With irreversible logic all the energy delivered by the source is eventually converted into heat. Hence, functional power, apart from leakage, will always be dissipated, making it worthwhile finding opportunities to reduce it to a minimum.

The functional power dissipation of a circuit node can be reduced by reducing the node transition-cycle activity factor, the clock-frequency and the transition-cycle energy.

Reduction of the node transition-cycle activity factor, the clock frequency or the node capacitance reduce the functional power dissipation linearly. The total physical node capacitance is reduced by scaling the transistor and wiring dimensions as well as the application of new isolating materials. The wiring capacitance does not scale as aggressively as the transistor capacitance, since the average wire length scales less rapidly compared to transistor dimensions.

The impact of static and dynamic voltage scaling on the functional power dissipation is significant, because of the quadratic relationship.

For new technology generations quality factors of resonant charging techniques will have to be reduced to increase circuit speeds, making these techniques inefficient.

5

REDUCTION OF PARASITICAL
POWER DISSIPATION

In chapter 3 the parasitical power dissipation, consisting of leakage power and short-circuit power, has been discussed. The parasitical power is dissipated without attributing to the actual changes of the internal circuit states. In this chapter solutions to reduce the leakage power and the short-circuit power will be discussed in section 5.1 and 5.2, respectively. In battery operated applications power dissipation caused by weak-inversion currents forms a large part of the parasitical power dissipation. Therefore, the last section will elaborate on the need for weak-inversion current reduction techniques.

All formulas and examples in this chapter will be defined for n-channel MOS transistors.

5.1 Leakage power dissipation

The leakage power dissipation is device related parasitical power. It is dissipated even when the circuit is idle and becomes dominant in standby modes, since no information is being processed and therefore the functional and the short-circuit powers are zero. Device scaling causes channel lengths, oxide thicknesses and threshold voltages to decrease 30% per generation, resulting in increased leakage power. In section 3.3.1 the leakage power has been divided into three sub-groups:

- channel leakage current;

- diode leakage current;

- gate leakage current.

Power reduction techniques for channel, diode and gate leakage currents will be discussed in section 5.1.1 through 5.1.3, respectively.

5.1.1   Channel leakage current

In section 3.3.1.1 it has been discussed that channel leakage currents consist of drain-source currents, which are present even when gate-source voltages are zero. Three effects contributing to channel leakage currents have been distinguished:

- weak-inversion current;

- drain-induced barrier lowering (DIBL) current;

- channel edge current.

Reduction techniques are discussed in the following sections.

5.1.1.1   Weak-inversion current

In section 3.3.1.1 the following expression for the weak-inversion current has been determined as a function of the threshold voltage (equation 3.42 on page 28):

$$I_{ds} = I_0 \exp\left(\frac{U_{gs} - U_{th}}{nU_T}\right)\left(1 - \exp\left(\frac{-U_{ds}}{U_T}\right)\right) \qquad (5.1)$$

Equation 5.1 indicates an exponential relationship between the gate-source potential and the weak-inversion current. Therefore, a weak-inversion current is still present for a gate-source potential equal to zero. During standby mode this weak-inversion current becomes dominant. For constant U_{ds} the weak-inversion current is influenced by five factors:

- gate-source voltage (U_{gs});

- threshold voltage (U_{th});

- weak-inversion slope ($S_{wi} \propto n$);

- device geometries (W_{ch}, L_{ch}, t_{ox}).

- temperature (T).

All these factors will be discussed successively.

Negative gate-source potentials can reduce weak-inversion currents, as depicted in figure 5.1(a) and (b). Since I_{ds} depends exponentially on U_{gs} small negative bias potentials reduce the weak-inversion current considerably. However, the negative gate-source potential cannot be increased unlimited by pulling the gate potential below the source potential, figure 5.1(a), since GIDL currents and, for ultra thin gate oxides, gate leakage currents will increase tremendously and dominate weak-inversion currents, as has been discussed in section 3.3.1.2

and section 3.3.1.3, respectively. Increasing the source potential with respect to the gate reduces besides the weak-inversion current DIBL as well, since the drain-source potential is decreased, see figure 5.1(b). It should be noted that the noise margins are reduced for lower drain-source potentials.

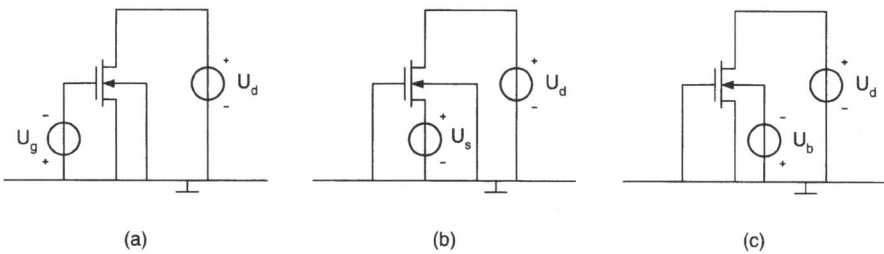

Figure 5.1. (a) Gate, (b) source and (c) substrate biasing techniques.

The threshold voltage has been defined in section 3.3.1.1 in terms of the strong-inversion surface potential (equation 3.38 on page 27):

$$U_{th} = U_{th0} + \gamma \left(\sqrt{\phi_0 + U_{sb}} - \sqrt{\phi_0} \right) \qquad (5.2)$$

Increasing the threshold voltage results in an exponentially decreasing weak-inversion current. The threshold voltage can be adjusted technologically or electrically. Technological adjustment of the threshold is achieved by adaptation of the channel-implant doping concentration, i.e. the substrate doping concentration N_{sub} at the channel surface, which influences the threshold voltage via ϕ_0 and γ, see equations 5.2. Electrical adjustment of the threshold voltage, also called substrate biasing, can be achieved via the source-bulk potential, see figure 5.1(c) and equation 5.2.

The weak-inversion "slope", is given by:

$$S_{wi} = nU_T \ln 10 \qquad (5.3)$$

Besides temperature, the weak-inversion slope depends on the slope factor, n, via the ratio between C'_d and C'_{ox}. For the slope factor holds:

$$n = 1 + \frac{C'_d}{C'_{ox}} \qquad (5.4)$$

The minimum value for S_{wi} equals $60mV$ per decade and is reached for $n = 1$.

The weak-inversion current can be reduced by decreasing S_{wi}. Increasing the oxide capacitance and/or reducing the depletion capacitance decreases S_{wi}. The weak-inversion slope has not changed significantly as technology advances, mainly because the gate oxide thickness has been scaled down but substrate doping profiles have been improved. The latter is necessary to reduce short channel effects, but increases the depletion capacitance.

The influence of *device geometries* on the weak-inversion current is included in an effective threshold voltage in the long channel approximation of equation 5.1 as has been discussed in section 3.3.1.1. The increase in weak-inversion currents for shorter, or smaller STI-isolated channels is modelled by lower effective threshold voltages. This short-channel effect is called U_{th} roll-off. To minimize this effect, the transistor lateral-to-vertical aspect ratio must be preserved from one technology generation to the next. This requires the channel length and width, the gate oxide thickness, the junction depth and the depletion layer width to scale down proportionally per generation.

Reducing the chip temperature reduces the weak-inversion slope, i.e. the slope becomes steeper, and increases the threshold voltage leading to an exponential reduction in the weak-inversion current, as discussed in section 3.3.1.1. Since in standby periods the overall on-chip power dissipation reduces, the chip temperature drops, reducing all temperature-related leakage currents.

5.1.1.2   Drain-Induced Barrier Lowering current

Drain induced barrier lowering is more profound in short-channel transistors and for increased drain bias, as has been discussed in section 3.3.1.1. DIBL can be reduced by:

- reducing depletion layer widths;

- reducing the drain-source potential.

Depletion layer widths around source and drain and in the channel region can be reduced by increasing the substrate doping concentration and reducing the source and drain junction depths. Therefore, C'_d values increase while surface potentials decrease and thus the injection rate of carriers from the source. A drawback, however, is the reduced grip of the gate on the channel region and the increased slope factor. Decreasing the gate oxide thickness or the use of high-permittivity gate oxides solves this problem. It should be noted that leakage through the gate oxide by direct band-to-band tunneling limits physical oxide thickness scaling, unless high-permittivity gate oxides are used, and reducing source and drain junction depths degrades drive currents. It should be noted that for junction depths on the order of the diffusion length pn-junction leakage increases rapidly. Higher source and drain doping concentrations lower the parasitical resistance, but in combination with the increased substrate doping concentrations could lead to band-to-band tunneling in degenerated pn junctions, causing excessive leakage currents to the substrate.

Reducing the drain source potential reduces leakage caused by DIBL, because the surface potential along the channel is reduced introducing less barrier lowering at the source side, as explained in section 3.3.1.1 and indicated in figure 3.16 on page 34. When applying a bias source as indicated in figure 5.1(b) the leakage current is reduced by the negative gate source potential and by reduced DIBL.

5.1.1.3 Channel edge current

Reducing transistor isolation areas is as important as shrinking device areas to boost device densities. From the 0.25μm CMOS technology generation abrupt shallow trench isolation technology has been applied. Channel edge currents can be reduced by introducing a bit more gradual transitions between active and isolation areas. In section 3.3.1.1 it has been discussed that the height of the isolation with respect to the silicon surface is the most important parameter considering channel edge currents, because it influences the control of the gate on the channel edge. The parameters influencing the channel edge current, presented in section 3.3.1.1 and figure 3.18 on page 36 are:

- the isolation height h with respect to the silicon surface;

- the transition angle θ_t;

- the corner radius r;

- the oxide fixed charge density N_{ss};

- the channel doping concentration N_{sub};

- the oxide thickness t_{ox}.

Significant reduction in channel edge currents can be obtained by the use of an overflowing oxide, compared to the recessed oxide situation [24].

Other parameters reducing the edge current include the transition angle and corner radius. A transition angle between the active and the isolation region of $20°$ reduces the parasitical edge current by one decade, whereas a corner radius of 100nm provides for half a decade reduction [24].

The fixed charge density (N_{ss}) in the gate insulator has opposing effects on the NMOS and PMOS transistors. Since these fixed charges always have a positive nature, they reduce the effective threshold of the edge transistors for NMOS and increase it for PMOS transistors.

Decreasing the channel doping concentration N_{sub} and the gate oxide thickness t_{ox} reduces the drop in effective threshold voltage of the edge transistor, compared to the main channel transistor. This equalizes edge currents and main channel currents.

Although an overflowing trench oxide yields the best results, it introduces step coverage when depositing the polysilicon gate. Step coverage provides for stress and the risk of cracks in layers caused by level differences in lower layers. Accordingly, one should aim for coplanar active/isolation areas.

An STI integration technology used by IBM for their $0.18\mu m$ generation uses two chemical mechanical polishing (CMP) steps with a reactive ion etching (RIE) step in between to make the active and isolation areas coplanar [26]. A silicon nitride (SiN) isolation mask defines the trenches, which are etched by RIE. The lowest stress, interface charge trap states and off-state leakage currents are obtained when the sidewalls of the trenches are cleaned and oxidized, i.e. liner oxidation, in a wet oxidation scheme at a temperature of 1150°C. Thereafter the trenches are filled with a chemical oxide (TEOS), see figure 5.2(a). In

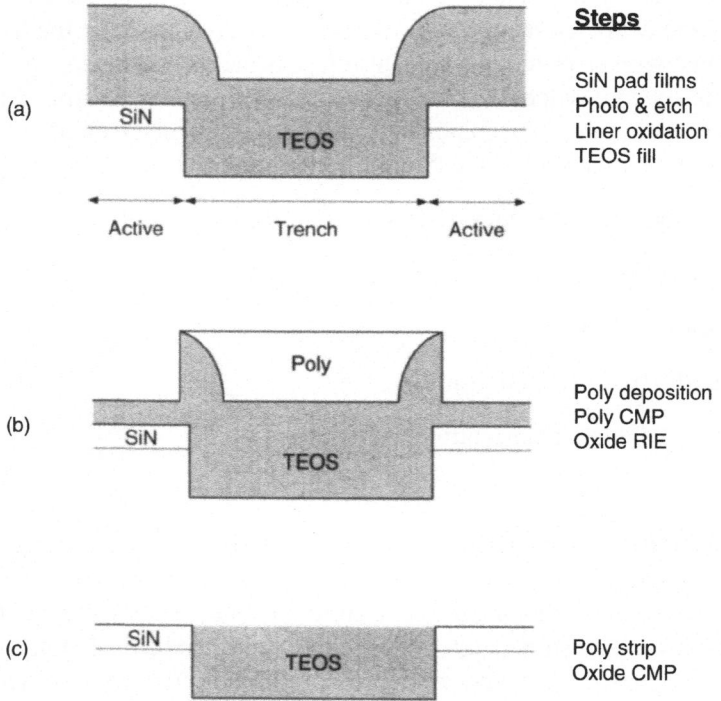

Figure 5.2.   Basic steps of the double-CMP process necessary to form a coplanar active/isolation area transition [26].

the next step a polysilicon layer is deposited on top of the trench fill TEOS, followed by a polysilicon CMP. This CMP step exposes the oxide on top of silicon over the active areas, but leaves polysilicon in place on top of the field oxide over the trenches. Subsequently, the RIE step etches the TEOS oxide over the active areas. Since polysilicon is a self-aligned etch stop for the oxide RIE, the TEOS humps at the edges of the trenches remain intact, see figure 5.2(b). Next the polysilicon is stripped and an oxide CMP removes the remaining TEOS over the active areas, see figure 5.2(c). The TEOS humps prevent dishing, i.e. recessed oxide, of the relatively large field oxide areas during the oxide CMP planarization step.

5.1.2 Diode leakage current

The diode leakage current consists of source-bulk and drain-bulk pn-junction leakage currents. Diode leakage consists of reverse bias leakage and gate-induced drain leakage. Reduction of both components will be discussed in section 5.1.2.1 and 5.1.2.2, respectively.

5.1.2.1 Reverse bias leakage

The reverse-bias leakage current consists of a reverse saturation and a generation current component. The total reverse-bias current has been presented in section 3.3.1.2 on page 39 and can be described by:

$$I_{reverse-bias} = A_j \left(J_s + J_{gen} \right) \tag{5.5}$$

In general this equation shows that reduction of the pn-junction area A_j reduces both the reverse saturation and the generation current components. Both components will be examined successively to discuss which parameters are involved in reducing each of them.

The reverse saturation current density J_s, i.e. a diffusion current, presented in section 3.3.1.2 on page 41 equals:

$$J_s = qn_i^2 U_T \left(\frac{\mu_n}{L_n N_d} + \frac{\mu_p}{L_p N_a} \right) \tag{5.6}$$

Temperature has a dominant effect on J_s, since the intrinsic carrier concentration, n_i, is exponentially dependent on temperature. The decrease in carrier mobility caused by lattice scattering is negligible compared to the increase in n_i.

From equation 5.6 it can be seen that increasing either donor or acceptor concentration or both reduces J_s. This can be explained by the increased recombination probability of injected minority carriers and the decreased mobility, caused by increased impurity scattering, at higher doping concentrations. In controlling short channel effects for subsequent technologies the bulk doping concentration is increased continuously. Higher doping concentrations increase the built-in potential U_{bi}, see equation 5.7, and thereby the band bending.

$$U_{bi} = U_T \ln \left(\frac{N_a N_d}{n_i^2} \right) \tag{5.7}$$

Therefore, in highly doped junctions the valence and conduction band on either side of the junction might come sufficiently close together that under reverse bias conditions electrons may tunnel, apart from trap-assisted tunneling, directly from the valence band on the p-side into the conduction band on the n-side, culminating in significant leakage currents.

The generation current density J_{gen} presented in section 3.3.1.2 on page 41 equals:

$$J_{gen} = \frac{q n_i t_{dep}}{2\tau_l} \tag{5.8}$$

The width t_{dep} of the depletion layer also has been introduced on page 41 and equals:

$$t_{dep} = \frac{2K_{Si}(U_{bi} + U_{db})}{qN_a} \tag{5.9}$$

As can be seen from equation 5.9 the larger the reverse junction voltage, the deeper the depletion layer "grows" into the bulk, i.e. the least doped region, the larger the generation current. Therefore, smaller reverse-bias potentials and, to a certain extent, higher substrate dopes decrease generation currents. Since J_{gen} is a function of n_i, it also is quite temperature dependent.

5.1.2.2 Gate-induced drain leakage

Gate-induced drain leakage (GIDL) is a leakage current from drain to substrate, caused by high electric fields between gate and drain. In section 3.3.1.2 it has been discussed that these electric fields provide for a deep depletion layer in the drain, in which carrier pairs are generated thermally. For low electric fields carriers can tunnel via surface traps through the depletion layer to both drain and substrate. GIDL leakage can be decreased by reducing the width of the drain's deep depletion layer. This will reduce the volume in which carriers can be generated thermally, as well as the volume of surface traps within the depletion area. The latter decreases the trap-assisted tunneling component of the GIDL current.

The width of the deep depletion region and thereby GIDL can be controlled by:

- the gate insulator thickness;

- the junction depth;

- the doping concentration of the drain;

- the surface trap density;

- the potentials across the region.

Increasing the gate insulator thickness would decrease GIDL, since electric fields will be reduced. However, controlling short channel effects while scaling down device geometries requires aggressive scaling of the gate insulator as well. Therefore, increasing the gate insulator thickness is not an option. The application of new gate insulators with higher permittivity K_{ins} will increase C'_{ins} and therefore reduce short channel effects. However, GIDL is not reduced, while maintaining E_{ins} constant, since the depleted charge per unit drain area increases:

$$Q' = K_{ins}E_{ins} \qquad (5.10)$$

Reduction of the drain junction depth reduces the deep depletion volume and therefore GIDL. However, it should be kept in mind that reducing the junction depth degrades drive current.

Increasing the doping concentration of the drain will reduce the deep depletion volume and thereby GIDL. As has been discussed in the previous section, increased doping concentrations provide for band bending, which might lead to increased junction leakage caused by band-to-band tunneling. The use of Lightly Doped Drain (LDD) structures is not favorable either, since they cause large depletion volumes. However, LDD structures are used to reduce lateral fields and prevent carriers from being emitted into the gate insulator, i.e. hot carriers, thereby reducing the integrity of the insulation layer. Since hot carriers generate interface states [58] they will increase GIDL. Increased channel dopes can also increase GIDL, since they force the depletion region into the drain area.

Careful fabrication of the interface between the channel surface and the isolation region, i.e. gate oxide, will significantly reduce the number of surface traps. The number of interface traps can be reduced by a thermal anneal step [59]. Another solution is the use of hydrogen. The hydrogen atoms attach to the dangling bonds, i.e. defects in the lattice, at the silicon silicon dioxide interface and reduce the number of traps.

High electric fields across the deep depleted drain region increase GIDL, because they increase the width of the region and increase the tunneling probability. As can be seen from figure 5.3 reducing the drain gate potential decreases GIDL. The vertical bars indicate the ratio between the current from the upper and lower curve. Although the influence of the substrate electrode on the width of the depletion layer in the drain is less than that of the gate, the effect is not negligible. From figure 5.4 it can be seen that GIDL currents increase for larger U_{sb} potentials.

5.1.3 Gate leakage current

Ultra thin gate oxides are necessary for gate electrodes to keep grip on channel regions while decreasing transistor dimensions. However, decreasing gate oxide thickness below $1.5nm$ will introduce direct-tunneling leakage-current densities higher than $1A/cm^2$. Boron penetration from the polysilicon gate through the gate oxide into the channel region causes threshold and flat band voltage shifts, trapped oxide charges and mobility degradation [60]. Trapped charges in the gate insulator enable trap-assisted tunneling, which increases gate

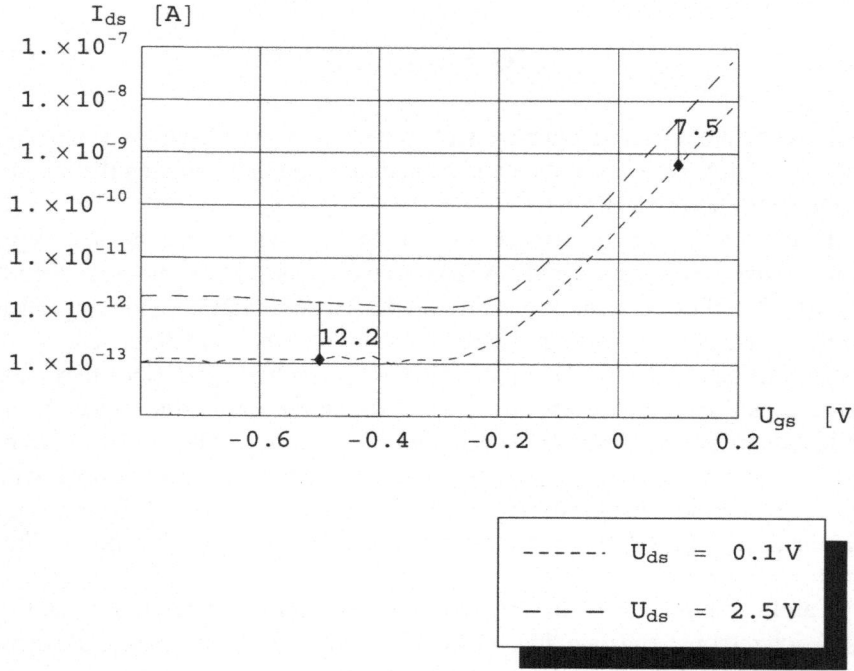

Figure 5.3.   Reduced GIDL currents for reduced drain biases for a 0.25μm NMOS transistor.

leakage currents. As has been discussed in section 3.3.1.3 two gate tunneling current components can be distinguished:

- gate-to-channel direct tunneling current;

- source and drain extension-to-gate overlap tunneling current.

In particular *source and drain extension-to-gate overlap tunneling currents* can be minimized by reducing the source and drain to gate extension areas, which is determined by the precision of the gate alignment.

Both tunneling components can be reduced by reducing the electrical field across the gate insulator. This can be achieved by:

- reducing bias voltages across gate insulator;

- increasing physical gate insulator thicknesses.

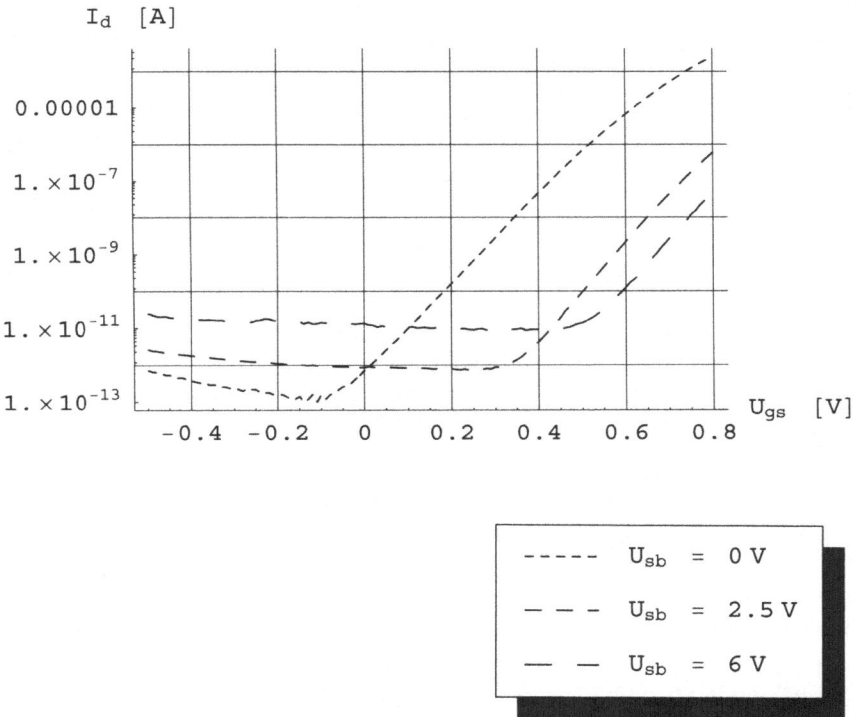

Figure 5.4. A 50μm wide and 0.25μm long transistor with 20 edges, with $U_d = 2.5$V, showing increased GIDL for increased U_{sb} biases.

Although reduction of bias voltages degrades transistor performance in active modes, it could be applied in standby modes.

To increase physical insulator thickness, without reducing the gate capacitance, can be accomplished by increasing the permittivity of the insulator material. At present two methods can be distinguished to create increased-permittivity gate insulator stacks:

- deposition of high-permittivity materials;

- implantation of radicals in the gate insulator.

Examples of high-permittivity materials [61] are titanium dioxide, TiO_2 (K_{TiO_2} ranges from 80 to 170), zirconium dioxide, ZrO_2 (K_{ZrO_2} ranges from 12 to 16) and silicon nitride, Si_3N_4 ($K_{Si_3N_4} = 7.9$). Titanium dioxide is deposited directly onto the crystalline silicon channel area [62, 63], whereas zirconium dioxide [64] and the silicon nitride are deposited on an oxynitride interfacial layer [65]. For all high permittivity materials a trade-off exists since the band gap, or more importantly the barrier height between gate and channel tends to decrease with increasing dielectric constant K_{ins} of the insulator [66]:

$$E_g \approx 20 \left(\frac{3}{2 + K_{ins}} \right)^2 \tag{5.11}$$

Reduced barrier heights are responsible for increased leakage currents by thermionic emission.

The oxynitride layer provides for a good interface between the silicon dioxide (grown on the channel area) and the zirconium dioxide or silicon nitride layer, which ensures a high channel mobility. The reported disadvantages of a TiO_2 layer are poor thermal stability [61], low barrier height and a substantially lower carrier mobility due to interface states compared to thermal oxide MOSFET's [62]. A disadvantage of the ZrO_2 stack is the large density of fixed negative charge [64], shifting the flat-band voltage and providing for traps. An advantage of the silicon dioxide/oxynitride/silicon nitride stack is the combination of an increased permittivity with an increased physical oxide thickness and a reduced boron penetration [65]. Moreover, the flat-band voltage shift compared to a thermal oxide MOSFET is negligible.

Creation of oxynitrides by implantation of nitrogen radicals via thermally enhanced remote plasma nitridation (TE-RPN) is a method to increase the permittivity and the physical thickness of a silicon dioxide layer [60, 67, 68]. First a thin SiO_2 basic layer is thermally grown onto the crystalline silicon channel region. Thereafter, this layer is exposed to a short, high-density, remote helicon-based nitrogen discharge. The final insulator thickness is larger then the basic layer. The increase in physical thickness is due to the nitrogen radical, N', which breaks the $Si - O$ bond to form a $Si - N$ bond while simultaneously releasing an O atom. The released O atom reacts with the silicon substrate to form a new $Si - O$ bond, and thus increasing the oxide thickness. Although the physical thickness of the insulation layer increased the effective, i.e. electrical, oxide thickness (EOT) reduced. The high nitrogen concentration piled up at the top surface of the oxide acts as a diffusion barrier, prohibiting boron from penetrating the channel region [68]. An important advantage of this technique is the simple fabrication process compared to other nitrided oxide or high-permittivity material.

Regarding technological developments the application of high-permittivity gate insulators to reduce gate leakage looks promising. However, there is still a long way to go before all technological problems have been solved.

5.2 Short-circuit power dissipation

Short-circuit power dissipation is circuit related parasitical power dissipation. Short-circuit power dissipation is characteristic only for static CMOS circuits, since a direct path between power supply lines can be created during switching transitions, as has been discussed in section 3.3.2. In equation 3.67 on page 47 an approximation of the short-circuit power dissipation for an inverter has been presented:

$$P_{short-circuit} = \alpha f_{clk} \frac{\beta \tau}{2^\sigma (\sigma + 1)} (U_{sw} - 2U_{th})^{\sigma+1} \qquad (5.12)$$

Means available to reduce the short-circuit power are:

- inhibit direct paths between power supply lines;

- reduce the node transition-cycle activity factor α;

- reduce the clock frequency f_{clk};

- reduce the logic signal swing U_{sw};

- increase the threshold voltage U_{th}.

A straightforward solution is to inhibit the direct path between the power supply lines during logic state transitions, as is the case for dynamic logic. An example of dynamic logic is the domino logic gate depicted in figure 5.5. However, the mainstream digital CMOS logic circuits are still static, because dynamic logic suffers from data reliability risks caused by external influences, e.g. alpha particles, discharging precharged circuit nodes.

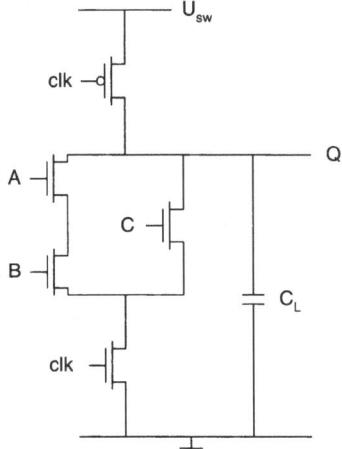

Figure 5.5. Domino logic gate, $Q = \overline{A \cdot B + C}$.

Solutions to reduce both α and f_{clk} have been presented in section 4.1 and 4.2, respectively and will be omitted here.

Reduction of the voltage swing and increase of the threshold voltage will reduce the short-circuit power. It can be seen from equation 3.67 that the short-circuit current can be made negligible for a power supply voltage smaller than the sum of the NMOS and PMOS threshold voltages, i.e. $2U_{th}$ in this case. However, in that case hysteresis will be introduced.

5.3   Need for weak-inversion current reduction

As has been discussed in chapter 4 many techniques tackling the functional power dissipation problem at the algorithm, system and circuit level have already been developed. This chapter introduced techniques to reduce the parasitical power dissipation. Solutions have been presented to reduce short-circuit power dissipation at system and circuit level, whereas mainly technological solutions have been introduced to reduce leakage power. Figure 5.6 gives an overview of the trends in the parasitical leakage power dissipation. Both weak-inversion and gate tunneling leakage currents will increase exponentially for new technology generations, and therefore become dominant compared to the diode leakage component. Although diode leakage currents decrease with the source and drain junction volumes, it should be noted that increased doping concentrations of source (drain) and bulk may lead to band-to-band tunneling currents. Moreover, junction leakage will also increase rapidly for junction depths on the order of diffusion lengths.

In section 3.4 the speed-performance leakage-power conflict has been introduced. It became clear that weak-inversion currents increase exponentially under constant field scaling conditions due to down scaling of threshold voltage levels. As a result power dissipation caused by weak-inversion currents becomes dominant during standby periods, because functional and short-circuit power dissipation are non-existent during these periods. Especially for battery operated applications weak-inversion current reduction is quite important.

Trends in parasitical leakage power dissipation:

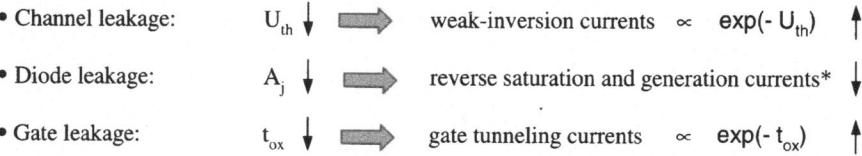

* I has been assumed that band-to-band tunneling is absent.

Figure 5.6.   Trends in parasitical leakage power dissipation.

Given a fixed amount of energy stored in a battery the maximum standby time is determined by the level of energy consumption caused by weak-inversion currents. Techniques reducing weak-inversion currents will be discussed in chapter 6.

5.4 Conclusions

In this chapter techniques reducing the leakage power, i.e. channel, diode and gate leakage, and the short-circuit power have been discussed.

Leakage power becomes dominant during standby periods, since the functional and short-circuit power dissipation are zero. Therefore, the presented reduction techniques mainly apply during these periods.

The weak-inversion and channel edge current components of the channel leakage, and the gate leakage become dominant compared to diode leakage, since both increase exponentially due to device scaling trends. Weak-inversion current reduction techniques become very important for battery operated applications, since weak-inversion currents determine the length of the standby time. Application of high-permittivity gate insulators to reduce gate leakage looks promising, although there is still a long way to go before all technological problems have been solved.

Short circuit power becomes negligible in static CMOS circuits for voltage swings lower as the sum of the NMOS and PMOS threshold voltages.

6

WEAK-INVERSION CURRENT REDUCTION

The need for weak-inversion current reduction is quite important for battery operated applications, since weak-inversion currents become dominant in standby periods. Therefore, weak-inversion currents will determine maximum standby times as has been discussed in section 5.3. This chapter will focus on weak-inversion current reduction techniques. Existing and new weak-inversion reduction techniques pass in review. To find new reduction techniques a classification has been developed, which enables a systematic search to find all possible weak-inversion current-reduction solutions. This classification is presented in the next section.

All formulas and examples in this section will be defined for n-channel MOS transistors. Analogous expressions can be found for PMOS devices.

6.1   Classification

The classification of weak-inversion current reduction techniques [69] depicted schematically in figure 6.1 is used as a guideline to find new and discuss existing weak-inversion current-reduction techniques.

At the highest level a distinction is made between two main weak-inversion current-reduction concepts:

■ power reduction with state retention;

■ power reduction without state retention.

Power reduction with state retention (left branch in figure 6.1) will be elaborated on in section 6.1.1 and could be applied in circuits which need to save circuit states during standby periods for seamless transitions to successive active periods, e.g. sequential logic. A solution used presently is substrate biasing. A

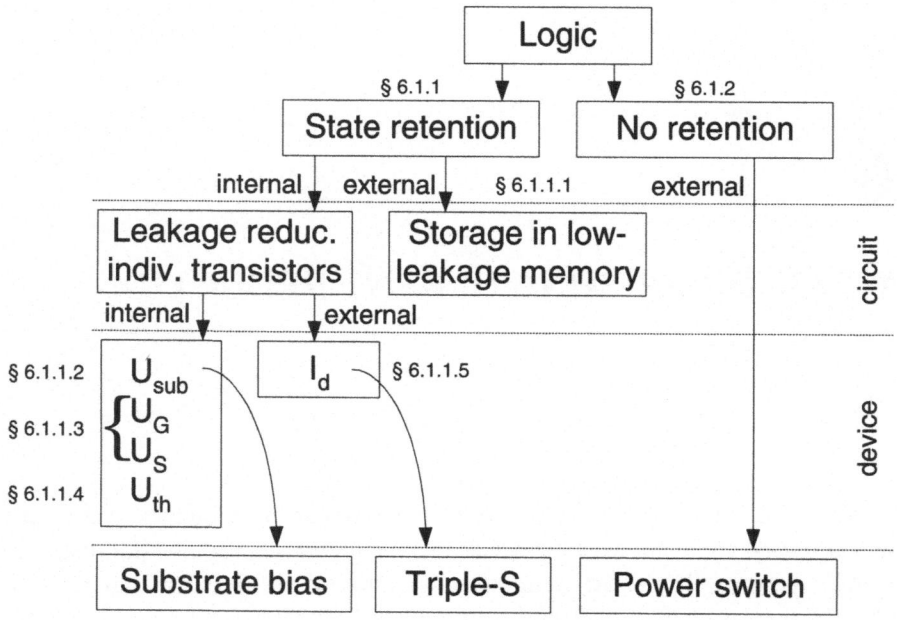

Figure 6.1. Classification of weak-inversion current-reduction techniques.

new technique is Triple-S and its principle will be introduced in this chapter, whereas practical results are presented in chapters 7 and 8. *Power reduction without state retention* (right branch in figure 6.1) is presented in section 6.1.2 and is applied in circuits which have no need to save circuit states during standby periods, i.e. combinatorial circuits. The power switch is already used in a wide extend.

6.1.1   Power reduction with state retention

The left branch of the classification tree (figure 6.1) indicates the class of techniques, which provide for weak-inversion current reduction and state retention during standby periods. A distinction has been made between circuit and device level. At the circuit level either the state logic itself can be made low-leakage by making the comprising devices, i.e. individual transistors, somehow low leakage (internal), or the state information can be stored in a separate optimized low-leakage memory and the rest of the leaky circuit is powered down (external). The latter option will be discussed in section 6.1.1.1. At the device level the leakage reduction of the memory itself (left branch) is again split into two subclasses. In the first subclass (internal), the voltages of the devices are controlled in such a way that the behavior of the transistors remains intact. The solutions belonging to this class will be discussed in section 6.1.1.2 through

6.1.1.4. In the second subclass (external), the current is determined externally to the device, i.e. the functioning of the device depends on components external to the device. In this subclass a new solution has been found, being Triple-S, which will be elaborated on in section 6.1.1.5.

6.1.1.1 Storage in intrinsically low-leakage memory

During standby periods system states can be saved in separate memory cells, which are intrinsically low leakage and subsequently the leaky memory can be powered down. Since speed performance is not an issue during standby periods, these memory cells could for instance consist of floating gate or high-threshold voltage devices. It should be taken into account that floating gates still allow a limited amount, i.e. $< 10^7$ times, of write cycles. For applications with relatively few but long standby periods, floating gates could be a boon.

System states can be stored also in flip-flops consisting of high-threshold-voltage transistors. For reasons of testability flip-flops are quite often connected in a scan chain. For this purpose every flip-flop is provided, besides the standard data input, D, with an extra test input, TI, which is enabled by the test enable input, TE, during test cycles. To form a chain, every flip-flop output, Q, is connected to the, TI, of the next flip-flop. The scan chain could be extended with flip-flops consisting of high-threshold voltage transistors, in which system states are stored during standby periods. A possible implementation is shown in figure 6.2. Since about 30% of chip area is occupied by flip-flops, this solution consumes worst case about 30% more chip area.

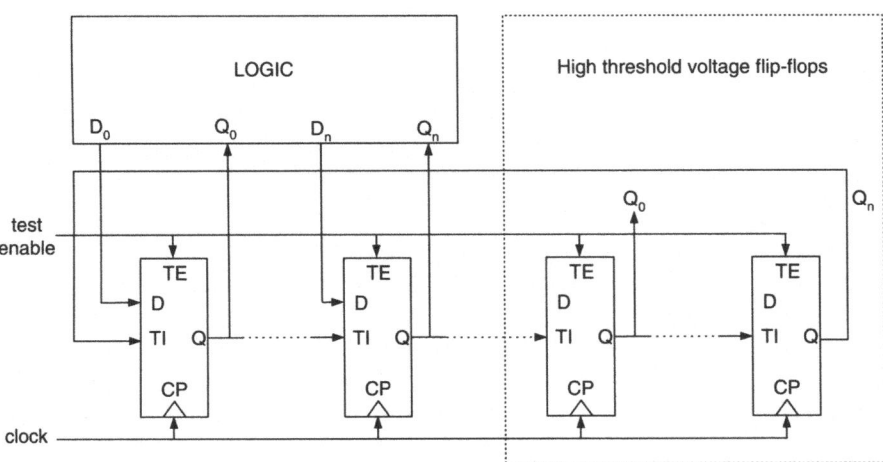

Figure 6.2. Scan chain extended with high-threshold-voltage flip-flops used to save system states and reduce leakage power during standby periods.

In stead of connecting them in a chain, flip-flops can be expanded with a separate local latch consisting of high-threshold voltage transistors, called balloon cell [70], in which data of the slave latch is stored during standby periods.

For memory cells a similar strategy could be followed. In that case the contents of some or all memory cells is stored in a separate memory consisting of high-threshold voltage transistors [71].

6.1.1.2 Substrate biasing

For a good speed performance low-threshold-voltage transistors are required, whereas for low weak-inversion leakage currents high-threshold-voltage transistors are required: the speed leakage-power conflict. Substrate biasing is a technique to shift the effective threshold voltage electrically via the bulk of the MOS transistor. A potential applied to the bulk electrode changes the surface potential and therefore the inversion layer charge density, causing a shift in the effective threshold voltage. Considering substrate biasing two strategies affecting the effective threshold voltage can be distinguished:

- mode controlled effective threshold voltage shift: VTMOS;

- state controlled effective threshold voltage shift: DTMOS.

The Variable Threshold voltage MOS (VTMOS) is a mode controlled substrate biasing technique [12, 57]. The effective threshold voltage is determined by the bias voltage applied to the bulk in the active or standby mode, see figure 6.3. A positive source-bulk potential will reduce the surface potential and thus increase the effective threshold voltage of the NMOS transistor. Hence, a static positive source-bulk potential applied during standby periods reduces weak-inversion leakage currents. In contrast during active periods a slightly negative source-bulk potential decreases the effective threshold voltage and therefore increases the saturation current and thus the speed performance. The latter technique is called substrate over-biasing [72]. It should be noted that the forward bias voltage should not exceed about $400mV$, otherwise the pn-junction will start conducting considerably, resulting in huge leakage currents. In VTMOS applications the common substrate is biased globally, i.e. independent of the state of individual transistors, see figure 6.4.

For the Dynamic Threshold voltage MOS (DTMOS) the gate electrode is connected to the substrate electrode (back gate), see figure 6.5(a). In fact it is a dual gate transistor although for bulk CMOS the back gate, i.e. the substrate, is much less effective compared to the front gate. Compared to VTMOS every DTMOS needs its own isolated well, to inhibit interaction between transistors [72, 73]. For conductive states both gates attract minority carriers into the channel, which inverts the channel more compared to a single gate MOS. For

DTMOS the weak-inversion slope becomes $60mV/decade$, i.e. $n = 1$, because gate and substrate are connected, see figure 6.6. When the DTMOS is in weak-inversion its drain current equals:

$$I_d = I_0 \exp \left(\frac{U_{gs} - U_{th}}{U_T} \right) \tag{6.1}$$

This is favorable for weak-inversion current reduction. For an equal saturation current, compared to a single gate MOS, the physical threshold voltage can be chosen higher, because the two gates have in total more grip on the channel. Compared to the single gate MOS equal threshold voltages result in larger weak-inversion current reductions, caused by the smaller n. It should be noted that saturation currents will have to be increased slightly, compared to the single gate MOS, caused by the increased gate capacitance. The risk of conducting forward biased bulk source pn-junctions lurks for voltage swings above approximately $400mV$. Larger voltage swings can be allowed by placing a reverse biased diode in between the gate and the substrate electrode, see figure 6.5(b).

The effectivity of substrate biasing will reduce for technology generations to come, as will be discussed in the next chapter.

6.1.1.3   Source and gate biasing

Biasing the source of a non-conducting MOS transistor during standby periods reduces weak-inversion currents, through the gate-source, the drain-source and the source-bulk potential. Figure 6.7(a) shows the principle of source biasing applied to a non-conducting NMOS transistor. In this case weak-inversion currents are reduced by the negative gate-source potential. On top of that the lowered drain-source potential and the body effect (lowered source-bulk potential) reduce the weak-inversion current further. The bias source signal has to be a function of the circuit mode, i.e. standby or active, and the state of the transistor, i.e. conducting or non-conducting. Activating bias sources to both conducting and non-conducting stages during standby modes would affect the

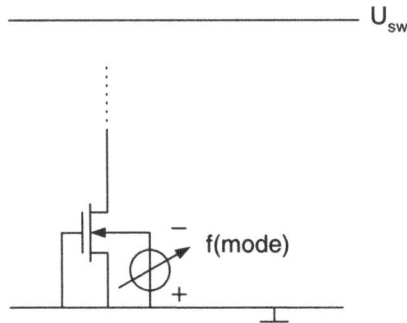

Figure 6.3.   Variable Threshold voltage MOS (VTMOS).

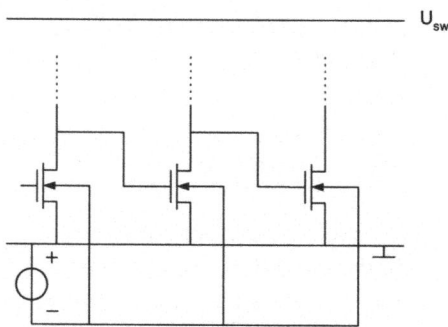

Figure 6.4. Biasing the common substrate of VTMOS transistors for controlling the effective threshold voltage.

voltage swing, as all source potentials are changed. Hence, the source biasing technique would degenerate to voltage scaling. Moreover, the bias source should have a low impedance, otherwise circuit performance is degenerated.

Biasing the gate of a non-conducting MOS transistor during standby periods reduces weak-inversion currents, through the gate-source potential. Figure 6.8 shows the principle of gate biasing applied to an NMOS transistor. In this case weak-inversion currents are reduced by the negative gate-source potentials. The gate can be biased independently of the state of the individual transistor, since the gate potential of the next stage is not influenced. The bias source should be floating and have a low impedance, otherwise circuit performance is degenerated. Since gate biasing increases the drain-gate potential above the voltage swing, by lowering the gate potential of the non-conducting device, GIDL currents are increased.

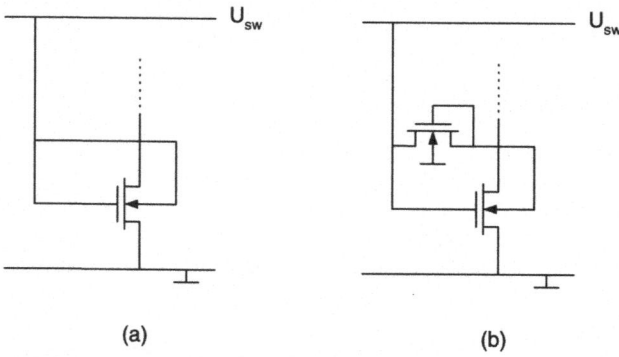

(a)   (b)

Figure 6.5. (a) Dynamic Threshold voltage MOS (DTMOS), (b) improved DTMOS.

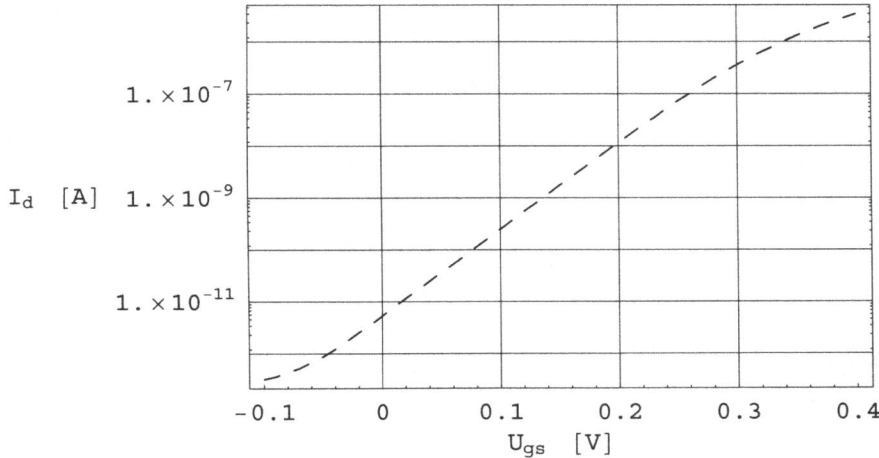

Figure 6.6. Weak-inversion slope of $60mV/decade$ for a DTMOS transistor. $W_{ch} = 5\mu$m, $L_{ch} = 0.25\mu$m, $U_{dd} = 2.5V$

6.1.1.4 Fixed high threshold voltage

Increasing the threshold voltage technologically results in an exponential decrease in weak-inversion currents. Of course saturation currents also reduce, resulting in decreased speed performance. Within reason this can be compensated by increasing transistor width. However, for thresholds reaching signal swings widths become infinite. In section 4.3.3.1, about static voltage scaling, it has been discussed that non-time-critical paths, within a circuit block, can be made time critical by the assignment of minimum signal voltage swings. Instead of reducing the signal voltage swing, transistors with a fixed high threshold

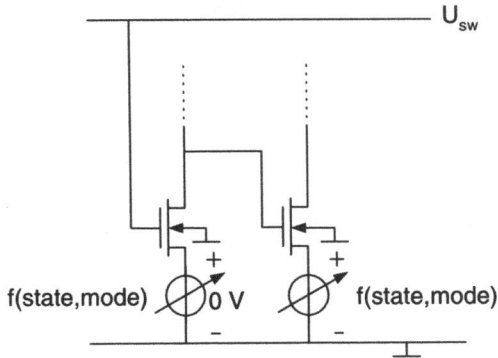

Figure 6.7. The principle of source biasing.

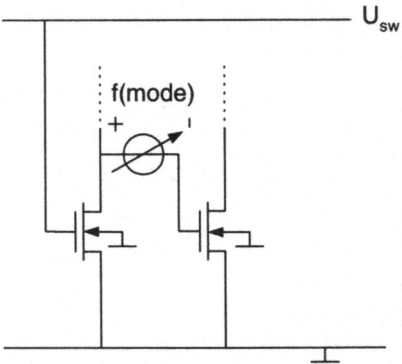

Figure 6.8.   The principle of gate biasing.

voltage are introduced in non-time-critical paths [14]. The application of fixed high threshold voltage transistors in external flip-flops and memory cells has already been elaborated on in section 6.1.1.1.

6.1.1.5   Triple-S Technique

In addition to adapting the weak-inversion current behavior of a leaky device internally, as has been discussed in section 6.1.1.2 through 6.1.1.4, the weak-inversion current can be determined by an external device also. Therefore, the external device has to be connected in series with the leaky device, as indicated in figure 6.9(a). In the standby mode the series device must be low leakage,

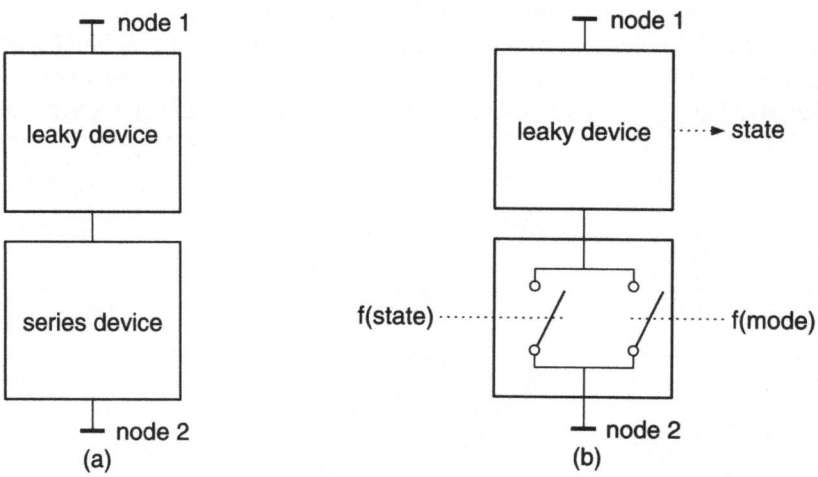

Figure 6.9.   (a) Series device determining the current through the leaky device, (b) The principle of Triple-S.

so its impedance should be higher than the impedance of the non-conducting leaky device. However, state retention may require the series device to have a low impedance in order to sustain a conducting path between the leaky device and node 2, e.g. a supply line. Accordingly, the impedance of the series device must depend on the state of the leaky device, i.e. whether it is conducting or not. In active mode the series device must possess a high speed performance, and therefore should always have a low impedance. Hence a series device is required whose impedance can be adapted as a function of both the operation mode and the state of the leaky device. These two functions are united in the *Smart Series Switch* [74, 75], or Triple-S for short, see figure 6.9(b). With Triple-S, each of the two functions is performed by a *separate switch*: one switches as a function of the operation mode, whereas the other switches as a function of the state of the leaky device. The two switches have to be connected in parallel, because the impedance of the series device has to be determined by the smallest of the two: the state switch must be able to overrule the mode switch in the standby mode for the sake of state retention, while the mode switch always overrules the state switch in the active mode for the sake of the speed performance. Accordingly, Triple-S solves the speed- performance leakage-power conflict.

Figure 6.10 shows a basic circuit implementation of Triple-S. The smart

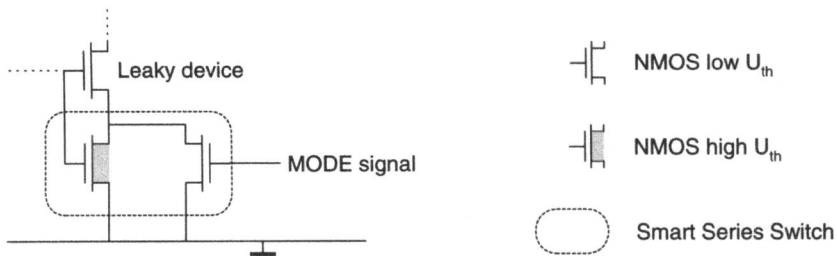

Figure 6.10. Basic implementation of the Triple-S technique.

series switch is connected in series with the "leaky device". This switch consists, in this implementation, of a parallel connection of a high and low-U_{th} device. When the circuit needs to be in the active mode, i.e. a good speed performance is important, the low-U_{th} mode switch is made conductive by means of the global mode signal. In this way the source of the "leaky device" is connected to the supply line and normal operation is possible. In standby mode the low-U_{th} mode switch is switched off. This is done by slightly reverse biasing the gate-source port. This in order to get the low-U_{th} device low-leakage. However, state retention can demand that the source of the "leaky device" remains connected to the supply line. This is realized by the high-U_{th} device. Depending on the state of the "leaky device" it is conducting or blocking. When the "leaky-device" is conducting, then also the high-U_{th} device is conducting, realizing state retention. When the "leaky device" is not conducting, also the high-U_{th} device is not conducting. The leakage is determined by the high-U_{th} device (i.e.

the current is externally determined, see classification of figure 6.1). As this device has a higher threshold voltage, the leakage can be considerably lower.

By this method, the performance in the active mode and the performance in the standby mode can be designed independently. The number of decades current reduction can be set by choosing the difference between the high and low threshold voltage.

6.1.2   Power reduction without state retention

The right branch of the classification tree (figure 6.1) indicates the class of weak-inversion current-reduction techniques without state retention during standby periods. Since there is no necessity to retain circuit states the circuits are switched off completely by disconnecting them from the power supply lines. This can be implemented by a switch external to the device or circuit. The class for which state retention is not a necessity, but is provided implicitly is placed in the class of state retention.

Weak-inversion current reduction can be achieved by determining the transistor's current behavior externally, i.e. an external device switches off the circuits completely, because there is no necessity to preserve the system states. Since the circuit consists of devices, i.e. transistors, external to the circuit always means external to the devices. Figure 6.11 shows the principle of the externally determined current technique without state retention, i.e. the power switch. The

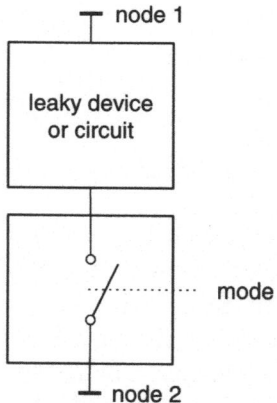

Figure 6.11.   Principle of the externally determined current technique without state retention: the power switch, for device and circuit application.

mode switch only switches as a function of the operation mode of the system, i.e. standby or active. The practical implementation of the mode switch yields a power switch in series with a leaky circuit or sub-circuit to disconnect power-supply line(s), node2, during standby periods. The performance of the mode switch is important, because it conducts during active modes and periods. Consequently, it must be able to source enough current from the power supply line to the series connected circuit, without seriously degrading speed performance. However, when it is non-conducting the impedance should be much higher than the impedance of the non-conducting leaky device or circuit, otherwise no weak-inversion current reduction is obtained. Since the saturation current reduces between linearly and quadratically with increasing threshold voltage, whereas the weak-inversion current reduces exponentially, an optimum transistor width can be found for a given physical threshold voltage and a reduction in speed performance. The procedure to determine the optimum width of the power switch W_{psw} is as follows [76]. Determine the current voltage relation, i.e. $I_{ds} - U_{ds}$ characteristic, of the power switch. From this the impedance times unit width, $R^*_{psw,on}$, can be deduced. Thereafter, the normalized delay of the circuit as a function of the supply voltage is determined. Given the allowed reduction in speed performance yields the voltage drop over the supply switch, ΔU_{psw}. Finally, determine the average current $\bar{I}$ through the circuit for a given operating frequency. Now the width of the power switch can be calculated from:

$$W_{psw} = \frac{R^*_{psw,on}}{\Delta U_{psw}} \bar{I} \tag{6.2}$$

The impedance $R^*_{psw,on}$ depends among other parameters on the threshold voltage of the power switch, $U_{th,psw}$. Another boundary condition for W_{psw} is the weak-inversion leakage current:

$$W_{psw} = \frac{I_{leak,max}}{I^*_0 \exp\left(\frac{U_{gs} - U_{th,psw}}{nU_T}\right)} \tag{6.3}$$

in which $I_{leak,max}$ and I^*_0 are the maximum allowable leakage current in standby periods and the leakage current per unit width of the power switch for $U_{gs} - U_{th,psw} = 0$, respectively. To fulfill both boundary conditions, the width calculated in equation 6.2 should be smaller than or equal to the width calculated in equation 6.3, yielding the following inequality:

$$U_{gs} \leq U_{th,psw} - nU_T \ln\left(\frac{R^*_{psw}\bar{I}I^*_0}{\Delta U_{psw}I_{leak,max}}\right) \tag{6.4}$$

This inequality implies that for widths of the power switch which suffice speed performance conditions for a given threshold voltage, the leakage can be reduced by the gate-source potential to reach the target leakage $I_{leak,max}$

In principle all options mentioned in section 6.1.1.2 through 6.1.1.4 are available to implement power switches:

- substrate biasing [77];

- source or gate biasing [78];

- fixed high threshold voltage [70].

6.2 Conclusions

A classification has been developed to find all possible weak-inversion current-reduction techniques systematically. This classification yields a new standby leakage reduction technique: the Smart Series Switch, or Triple-S for short. The classification distinguishes two mainstream leakage reduction techniques:

- power reduction with state retention;

- power reduction without state retention.

The first category includes two options. The first option offers the storage of system states in external memory cells, which are intrinsically low-leakage, like floating gates, or flip-flops or SRAM cells consisting of high-threshold voltage transistors. The second option involves the reduction of leakage currents of individual transistors. Several options to reach this have been discussed. Triple-S solves the speed-performance leakage-power conflict by implementing two separate parallel switches; the state dependent switch determines the impedance during standby modes, whereas the mode dependent one determines the impedance during active modes.

Leakage power reduction without state retention comes down to a power switch between the circuit and the power supply line. In principle the power switch can be made low-leakage during standby periods by application of substrate, gate or source biasing techniques, or fixed high-threshold-voltage transistors. The on resistance of the power switch determines the performance of the leaky device or circuit in active mode.

7

EFFECTIVENESS OF WEAK-INVERSION CURRENT REDUCTION

The effectiveness of energy reduction techniques is best observed when charge or energy delivered by the power supply source is considered, instead of comparing static leakage currents or power in standby and active periods (see figure 2.3 for definitions). Time aspects need to be taken into account. Addition of circuitry enabling weak-inversion-current reduction brings extra power dissipation with it. Therefore, the effectiveness and usefulness of leakage-power reduction techniques for specific applications is best determined in terms of the break-even standby time. The break-even standby time indicates a minimum standby time, required before power reduction techniques provide for any gains. In section 7.1 the impact weak-inversion current-reduction techniques have in general on break-even standby times is discussed. Section 7.2 considers the effectiveness of three specific (practical and theoretical) weak-inversion current reduction techniques.

7.1 General effectiveness

Reduction of weak-inversion currents can only become effective during standby periods, because the functional and the short-circuit power dissipation are non-existent. It has been assumed that functional power is much larger than the weak-inversion leakage power. Therefore, for each process frequency and voltage swing on the one hand and threshold voltage on the other hand should be properly balanced. A minimum or break-even standby time exists, because leakage reduction techniques only become profitable when the energy saved during standby periods is greater than the extra energy dissipated, caused by the addition of the leakage reduction circuitry. In the ideal case, discussed in section 7.1.1, the energy stored in the system's state capacitors is assumed to remain unaffected during standby periods, i.e. is not dissipated by leakage currents. Further, the energy dissipated in the auxiliary circuitry providing leakage reduction is assumed to be zero. Section 7.1.2 elaborates on the additional influence the energy stored in the system has on the average standby time. The impact of auxiliary circuitry, providing leakage reduction, on average standby times is revealed in section 7.1.3.

7.1.1   Ideal case

In this section power reduction in standby periods under ideal conditions is regarded [79]. No extra energy costs exist that have to be compensated before making profit when the system switches from active to standby period. Consequently, as soon as the system switches to the standby period energy is saved. This means that the break-even standby time is zero. Thus, the energy saved, E_{saved}, by switching to the standby period only depends on the total standby time, $T_{standby,tot}$, see figure 7.1, and equals:

$$E_{saved} = T_{standby,tot} \left(P_{leakage,not\ reduced} - P_{leakage,reduced} \right) \qquad (7.1)$$

in which $P_{leakage,not\ reduced}$ and $P_{leakage,reduced}$ represent the standby leakage power without the application of standby-energy reduction techniques and the reduced standby power, respectively. As long as the total standby times are equal the mode-switching frequency, i.e. the frequency at which the active and standby periods alternate, has no influence on the energy-reduction effectiveness.

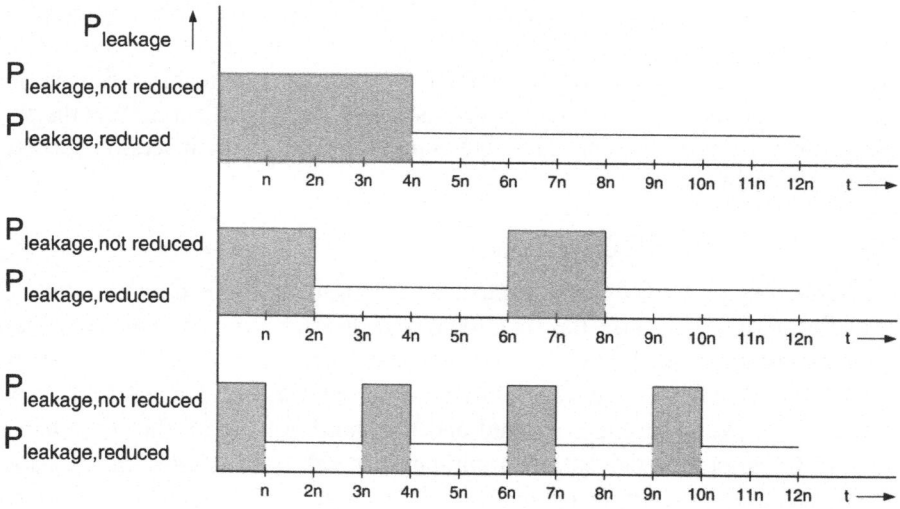

Figure 7.1. Equal ratios between standby and active times, 8/12, imply equal power reduction in ideal cases. The shaded parts indicate periods without current reduction. $P_{leakage,not\ reduced}$ and $P_{leakage,reduced}$ are the leakage powers in active and standby periods, respectively. Time unit n is arbitrarily chosen.

7.1.2 Dissipation of stored energy

In this section the ideal model is extended with the dissipation during standby periods of the energy stored in the system's capacitors [79]. Parts of digital systems in which there is no need for state retention during standby periods, e.g. combinatorial logic, can be switched off completely [75]. The moment the system enters the standby period, the power delivered by the supply source to the circuit is reduced immediately. However, system node capacitors will act as power supply sources and their charge will be drained by weak-inversion leakage currents of previous transistor stages, see figure 7.2. This energy (charge) will have to be replaced again during system recovery when switching to the active period. Consequently, leakage reduction only becomes effective when the standby period is greater than or at least equal to the time needed for all the energy stored in the circuit to dissipate, otherwise nothing will be gained. This time will be called the break-even standby time for stored energy T_{stdb,be_se}. The break-even point has been reached when the charges indicated by the shaded areas in figure 7.3 are equal (Note that the I_{source} axis is logarithmic). For the determination of this break-even standby time in this particular case the total system is modeled as an equivalent number of standard inverters having the equivalent standby and functional energy behavior. Each inverter consists of low-threshold-voltage transistors and the PMOS sleep transistor is a high-threshold-voltage transistor, see figure 7.2(a). During standby periods the

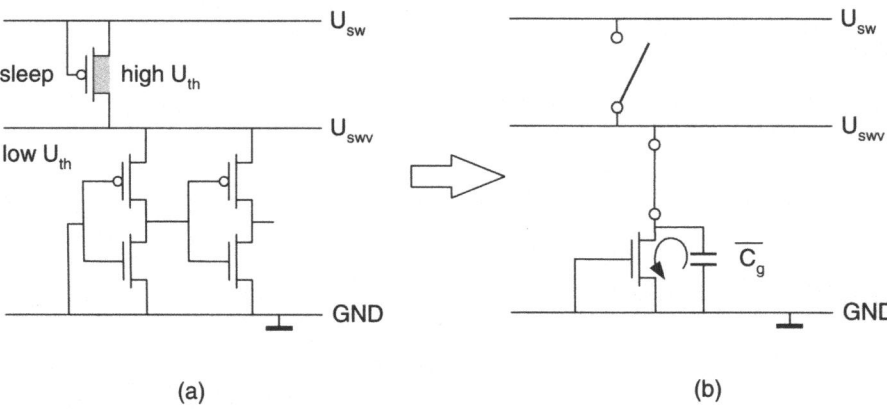

(a) (b)

Figure 7.2. Simplified inverter circuit during standby periods. The inverter's state has been chosen arbitrarily. The arrow indicates the low threshold weak-inversion current.

sleep transistor will be non-conducting, i.e. its weak-inversion current is many orders of magnitude lower than the weak-inversion current of the inverter's non-conducting NMOS transistor. The state of the considered inverter, i.e. the first one, has been chosen arbitrarily. The impedance of the inverter's PMOS transistor is considered negligible compared to that of the NMOS, because it is conducting, whereas the NMOS is blocking, see figure 7.2(b). In standby periods the weak-inversion current, $I_{wi,U_{th,low}}$, of the low-threshold-voltage NMOS transistor will discharge the average input capacitance, $\overline{C}_g$, of the next inverter. Constant (DIBL is neglected) and equal weak-inversion currents have been assumed for NMOS and PMOS transistors during the discharge of $\overline{C}_g$. The average time needed to drain the charge $\overline{C}_g\overline{U}_{sw}$, the average break-even standby time for stored energy, equals:

$$\overline{T}_{stdb,be\_se} = \frac{\overline{C}_g\overline{U}_{sw}}{I_{wi,U_{th,low}}} \tag{7.2}$$

Returning to the active period, the sleep transistor connects the inverter with the power supply line and the charge $\overline{C}_g\overline{U}_{sw}$ will be restored by the supply source again, see figure 7.3. If the charge indicated by the shaded areas is

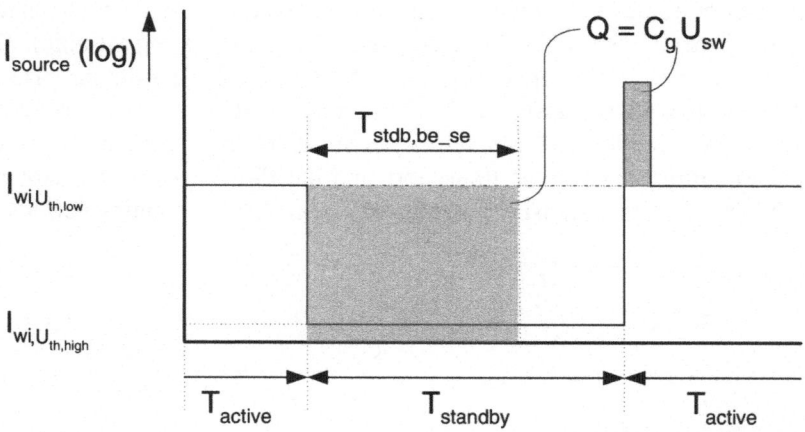

Figure 7.3.   Charge delivered by the supply source. Only the parasitical currents are drawn. Charges in the shaded areas are equal.

equal (I-axes is logarithmic) the break-even point has been reached. From the previous discussion it becomes clear that the minimum standby time and thus the mode-switching frequency is a function of the amount of charge stored in the system and the leakage current of the low-threshold-voltage devices. The total energy delivered by the supply source is only reduced when the average standby time equals at least $\overline{T}_{stdb,be\_se}$.

7.1.3   Overhead costs

Auxiliary circuitry, added to the system to reduce the standby energy dissipation, causes extra energy dissipation, i.e. overhead costs [79]. This section presents a break-even standby time determined only in terms of these overhead costs. The overhead costs can be divided into:

- a part due to an increase in the logic gate capacitance, which increases the functional energy dissipation;

- a fixed part due to, e.g. energy dissipation in charge pumps for generating biasing voltages.

The contribution of the fixed part per logic gate in the system is assumed to be negligible compared to the part due to the increase in the logic gate capacitance. Consequently, the increase in functional energy E_{func} is the only contribution to the overhead costs $E_{overhead}$. According to equation 4.1, page 53, and equation 4.8, page 56, the average functional energy for a circuit containing N circuit nodes (inverters in this case), an average node cycle activity factor α, an equivalent frequency $f_{clk,eq}$, a reversibility factor η, a load capacitance C_L and a voltage swing U_{sw} is expressed as:

$$\overline{E}_{func} = N\overline{\alpha} f_{clk,eq} \overline{(1-\eta)\,C_L\,U_{sw}^2 T}_{active} \qquad (7.3)$$

in which $\overline{T}_{active}$ is the average length of the active period in the standby mode (see figure 2.3 for definitions). For reversible logic the functional power dissipation depends on η, which equals 1 for quasi static processes. Maximum functional energy consumption is reached for irreversible processes, since $\eta = 0$. Therefore, this worst case situation will be regarded in particular. For irreversible logic the average overhead costs are expressed as a function of the functional energy and equal:

$$\overline{E}_{overhead} = \delta\overline{E}_{func} = \delta N\overline{\alpha} f_{clk,eq} \overline{C}_L \overline{U_{sw}^2} \overline{T}_{active} \qquad (7.4)$$

in which δ indicates the relative increase of the logic gate capacitance. Like in the preceding cases a break-even standby time can be determined by equating the energies of the shaded areas, see figure 7.4. It is assumed that the standby energy dissipation is reduced by a factor Red. Again, for the determination of the average break-even standby time for overhead costs, $\overline{T}_{stdb,be\_oh}$, the total system is considered as an equivalent number of standard inverters having an equivalent standby and functional energy behavior, see figure 7.5. Analogous to the case in the preceding section only the load of the first standard inverter needs to be considered. On average this load, $\overline{C}_L$, equals the total circuit capacitance divided by the total number of equivalent standard inverters, N, in which case the average capacitive load equals:

$$\overline{C}_L = \frac{C_{tot}}{N} = 2\overline{C}_g \qquad (7.5)$$

The extra functional energy consumed by one inverter equals:

$$\frac{\overline{E}_{overhead}}{N} = \delta\overline{\alpha} f_{clk,eq} \overline{C}_L \overline{U_{sw}^2} \overline{T}_{active} \qquad (7.6)$$

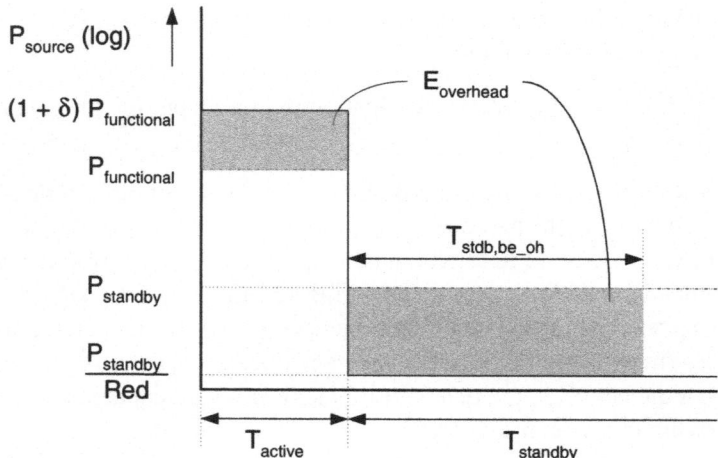

Figure 7.4.   Determination of break-even standby time.

The average saved energy during a standby period equals:

$$\overline{E}_{saved} = \left(I_{wi,U_{th,low}} - \frac{I_{wi,U_{th,low}}}{Red} \right) \overline{U}_{sw} \overline{T}_{standby} \approx I_{wi,U_{th,low}} \overline{U}_{sw} \overline{T}_{standby}$$

(7.7)

in which $Red \gg 1$. Equating equations 7.6 and 7.7, substituting equation 7.5, and solving for $T_{standby}$ results in the average break-even standby time for overhead costs:

$$\overline{T}_{stdb,be\_oh} = \frac{2\delta\overline{\alpha}f_{clk,eq}\overline{C}_g\overline{U}_{sw}^2\overline{T}_{active}}{I_{wi,U_{th,low}}\overline{U}_{sw}}$$

(7.8)

To reveal the dominance of loss through overhead costs over the loss of stored energy, equation 7.8 can be expressed in terms of the average break-even time

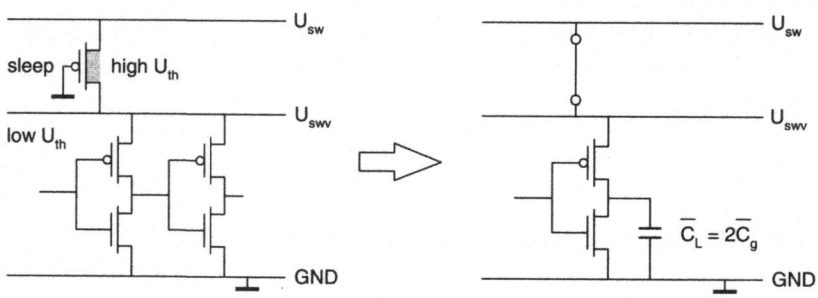

Figure 7.5.   Simplified loaded inverter in active periods.

in the case of stored energy:

$$\frac{\overline{T}_{stdb,be\_oh}}{\overline{T}_{stdb,be\_se}} = 2\delta\overline{\alpha}f_{clk,eq}\overline{T}_{active}\frac{\overline{U_{sw}^2}}{\left(\overline{U}_{sw}\right)^2} \qquad (7.9)$$

In case there are no parasitical transitions the node transition activity factor will be minimal ($\overline{\alpha} = 1$). Assuming a constant voltage swing yields:

$$\overline{U_{sw}^2} = \left(\overline{U}_{sw}\right)^2 \qquad (7.10)$$

When the addition of extra circuitry increases the total capacitance with only $1\%(\delta = 0.01)$ in a system where $f_{clk,eq} = 200MHz$ and $\overline{T}_{active} = 1ms$, it holds that:

$$\frac{\overline{T}_{stdb,be\_oh}}{\overline{T}_{stdb,be\_se}} > 1000 \qquad (7.11)$$

The above-mentioned values for δ, $\overline{\alpha}$, $f_{clk,eq}$ and $\overline{T}_{active}$ are optimistic predictions, in most cases they will be larger. Thus for most future systems the overhead costs are dominant and accordingly determine the minimum standby time for saving energy.

7.2 Technique-specific effectiveness

In section 7.2.1 through 7.2.3 the effectiveness of three weak-inversion current reduction techniques will be presented:

- substrate biasing;

- source and gate biasing;

- Triple-S technique.

7.2.1 Effectiveness of substrate biasing

Substrate biasing enables electronic control of the effective threshold voltage by application of a potential to the substrate of an MOS transistor, as has been discussed in section 6.1.1.2. In this section the development of the effectiveness of substrate biasing for future MOS technologies will be examined. Therefore, the amount of threshold voltage shift will be determined as a function of the maximum supply voltage per technology generation applied to the substrate. The dependency of the threshold voltage U_{th} on the source-bulk voltage U_{sb}, presented in section 3.3.1.1 on page 27, equals:

$$U_{th} = U_{th0} + \gamma\left(\sqrt{\phi_0 + U_{sb}} - \sqrt{\phi_0}\right) \qquad (7.12)$$

in which U_{th0} is the threshold voltage for zero source-bulk voltage, ϕ_0 is twice the difference of the intrinsic Fermi level and the Fermi level at the thermal equilibrium and γ is the body factor. For the body factor γ holds:

$$\gamma = \frac{\sqrt{2qK_{Si}N_{sub}}}{C'_{ox}} \tag{7.13}$$

N_{sub}, C'_{ox}, q and K_{Si} represent the substrate doping concentration, the gate capacitance per unit area, the electron charge and the permittivity of silicon, respectively. For future processes the trend is:

- C'_{ox} increases;

- maximum voltages reduce.

An increased C'_{ox} means a reduction of γ. Further, when maximum allowed voltages for a technology are reduced (due to the smaller dimensions), the maximum allowed bulk-source voltage also reduces. Hence, these two effects reduce the maximum possible variation in the threshold voltage.

Compensating the reduction of γ by increasing the doping level of the substrate is not an option. The dependency on the doping level is via a square root and thus a relatively large increase of the doping is required. Accordingly fully compensating the reduction of γ yields degeneration of drain-bulk and source-bulk junctions, resulting in strong band-to-band tunneling currents [80, 81].

Table 7.1 shows the trends in leakage current and leakage reduction effectiveness. These trends are depicted in figure 7.6.

Generation [μm]	0.5	0.35	0.25	0.18	0.13
U_{sb} [V]	5	3.3	2.5	1.8	1.5
γ [$\sqrt{V}$]	0.48	0.36	0.34	0.24	0.19
ϕ_0 [V]	0.66	0.66	0.64	0.70	0.70
$\Delta U_{th}[V]$	0.75	0.42	0.33	0.18	0.12
$I_{leakage}[A/\mu m]$	100f	42f	7p	400p	800p
$I_{leakage,reduced}[A/\mu m]$	0.03a	0.2a	0.2f	1p	30p

Table 7.1.   Maximum ΔU_{th} shift and weak-inversion current reduction by substrate biasing per generation.

From this plot it can be seen that the sub-threshold leakage current increases for next generations. This is of course due to the reduced threshold voltages. Further, it can be seen that the effectiveness of the substrate biasing reduces. Where for a 0.5 μm technology the reduction is about 6 decades, for a 0.13 μm

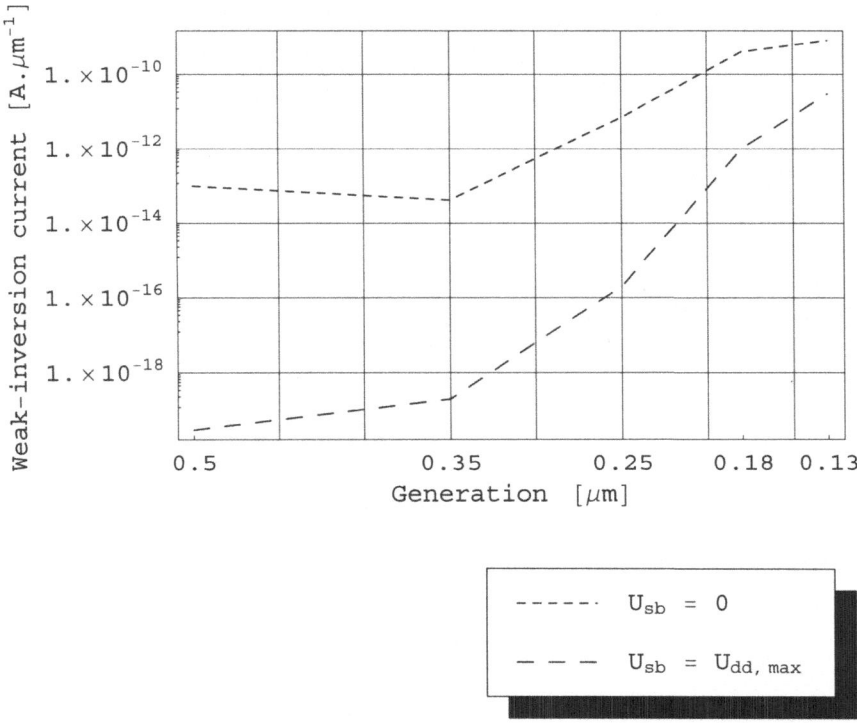

Figure 7.6. The trend in the sub-threshold leakage currents (upper line) and the reduced current when a substrate bias voltage equal to the supply voltage is applied (lower line).

technology this is only 2 decades [75]. Increasing substrate bias voltages above supply voltages will not reduce channel leakage currents, since GIDL currents increase, as has been discussed in section 5.1.2.2 and indicated in figure 5.4. Moreover, due to the global connection of substrate contacts it is more difficult to guarantee a uniformly biased substrate. This variation influences circuit speed performance.

So, the conclusion is that for future technologies other techniques need to be used to reduce sub-threshold leakage currents in order to gain standby time.

7.2.2 Effectiveness of source and gate biasing

Source biasing, discussed in section 6.1.1.3 reduces the weak-inversion current mainly through a negative gate-source potential, without introducing GIDL, see figure 7.7. On top of that the drain-source potential is reduced. The effect of the negative source-substrate potential on the body effect is neglected for present and future bulk CMOS processes, as has been demonstrated in the

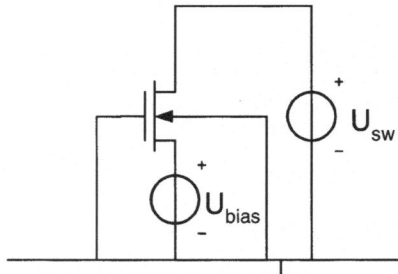

Figure 7.7.   Weak-inversion current reduction with source biasing.

previous section. For the weak-inversion current neglecting DIBL holds:

$$I_{ds} = I_0 \exp\left(\frac{U_{gs} - U_{th}}{nU_T}\right) \qquad (7.14)$$

It has been assumed that the reduced drain-source potential is still much larger than the Boltzmann voltage U_T. According to equation 7.14 for the weak-inversion current in case of source biasing holds:

$$I_{ds} = I_s \exp\left(\frac{-U_{bias}}{nU_T}\right) \qquad (7.15)$$

in which U_{bias} is the bias source in series with the transistor's source. Accordingly, $U_{gs} = -U_{bias}$ and I_s equals:

$$I_s = I_0 \exp\left(\frac{-U_{th}}{nU_T}\right) \qquad (7.16)$$

Figure 7.8 shows the weak-inversion current for a $0.25\mu m$ MOS transistor as a function of the source bias voltage. Clearly the exponential relationship is visible in the figure. For higher U_s the current is limited by the leakage of the measurement equipment. As discussed in section 6.1.1.3, in circuit applications the bias source potential should be a function of both the transistor state and the operation mode of the circuit, otherwise source biasing degenerates to voltage scaling.

Self-reverse biasing [82, 83] is a source biasing technique. The principle, indicated in figure 7.9, comes down to leakage current reduction in a transistor stack. The current through the transistor stack is reduced by the voltage drop across the lower transistor, which provides for the negative gate-source potential of the higher non-conducting transistor. Self-reverse biasing not always benefits from maximum leakage reduction, since the leakage reduction in a stack depends on the numbers and locations of the non-conducting transistors; there is no control over state and mode dependency of the bias source. For a

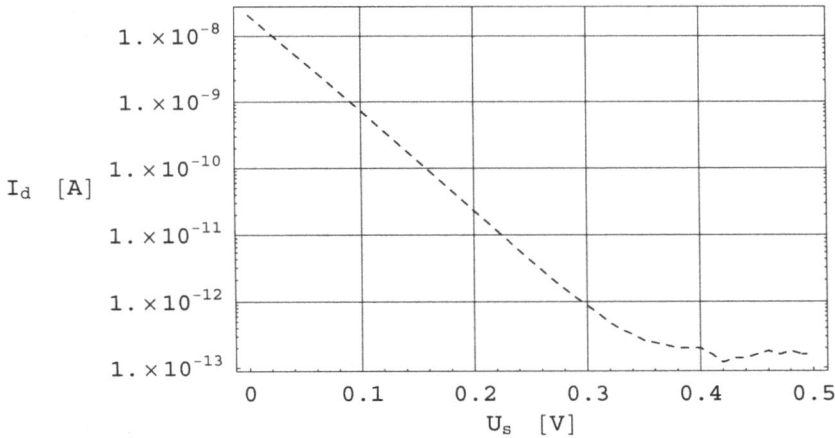

Figure 7.8. Weak-inversion current as a function of the source bias voltage for an NMOS transistor with $W_{ch} = 0.35\mu$m, $L_{ch} = 0.25\mu$m, $U_{drain} = 2.5V$ and $U_{th} = 200mV$, at $T_{amb} = 300K$.

weak-inversion slope of $80mV/decade$ and $U_{sw} \gg U_T$, and equal $\beta's$, it can be calculated that the drain-source potential U_{ds1} is approximately $20mV$. This will reduce the weak-inversion current about two times, as can be seen from figure 7.8.

For gate biasing weak-inversion current reduction is achieved only by a negative gate-source potential, see figure 7.10. Disregarding the effects caused by GIDL and gate leakage the weak-inversion current in case of gate biasing equals:

$$I_{ds} = I_s \exp\left(\frac{-U_{bias}}{nU_T}\right) \qquad (7.17)$$

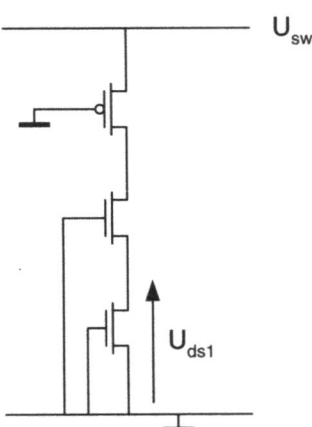

Figure 7.9. Weak-inversion current reduction with self reverse-biasing.

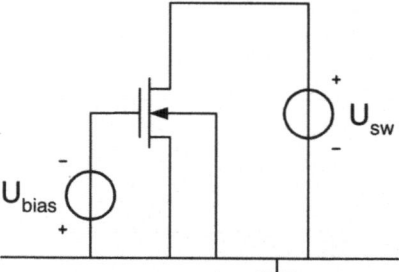

Figure 7.10.   Weak-inversion current reduction with gate biasing.

Figure 7.11 shows the weak-inversion current for a 0.25μm MOS transistor as a function of the gate bias voltage. Again the exponential relationship is visible in

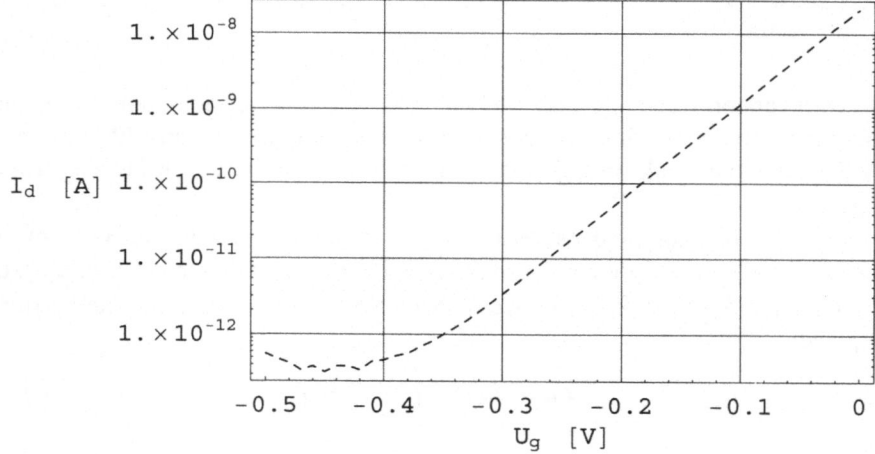

Figure 7.11.   Weak-inversion current as a function of the gate bias voltage for an NMOS transistor with $W_{ch} = 0.35\mu$m, $L_{ch} = 0.25\mu$m, $U_{drain} = 2.5V$ and $U_{th} = 200mV$ at $T_{amb} = 300K$.

the figure. For higher U_g the current is limited by the leakage of the measurement equipment. Implementing gate biasing in circuit applications presents practical problems, since a low impedance floating voltage source has to be connected in series with each gate in order not to decrease speed performance. No practical solutions have been found yet to implement this source efficiently.

Practical application of source biasing in power switches to reduce weak-inversion leakage currents is straightforward, since the state of these switches is always the same, i.e. non-conducting. Gate biasing could also be applied to power switches, however, it should be noted that GIDL and gate leakage currents will increase for future technologies.

7.2.3 Effectiveness of the Triple-S technique

In section 6.1.1.5 the principle of the Triple-S technique was presented and it has been discussed that it enables independent design of weak-inversion current reduction in standby periods and speed performance in active periods. Figure 7.12 shows a latch comprising Triple-S inverters. In principle the standby weak-

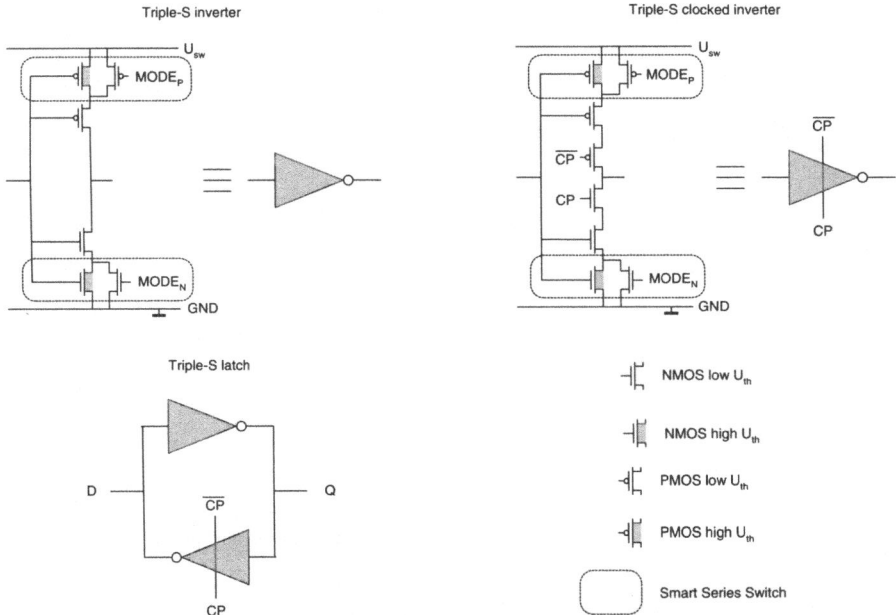

Figure 7.12. A latch comprising Triple-S inverters.

inversion current level is determined by the high-threshold-voltage state transistor. This is only true when the low-threshold voltage mode transistor can be switched off as good as the state transistor. The speed performance in the active periods is roughly reduced by a factor of two compared to the situation without the smart series switch, caused by the extra capacitive loading of the state switch. The Triple-S latch of figure 7.12 has been processed in a $0.25\mu m$ dual-U_{th} process. The low and high threshold voltage equal approximately $200mV$ and $600mV$, respectively. For this pair of threshold voltages the leakage will be comparable to processes of $0.13\mu m$ and beyond. In order to be able to measure the reduced leakage currents, i.e. reasonably above $1pA$, 5000 of these identical latches have been connected in parallel. All low threshold-voltage transistors have a W_{ch}/L_{ch} ratio of $0.35\mu m/0.25\mu m$, whereas the high-threshold voltage transistors have, to diminish U_{th} roll-off, a W_{ch}/L_{ch} ratio of $0.35\mu m/0.35\mu m$. The mode switch can be gate or source biased.

Figure 7.13 indicates the measured supply current, I_{dd}, i.e. the weak-inversion leakage current, as a function of the $MODE_P$ bias voltage, which is applied

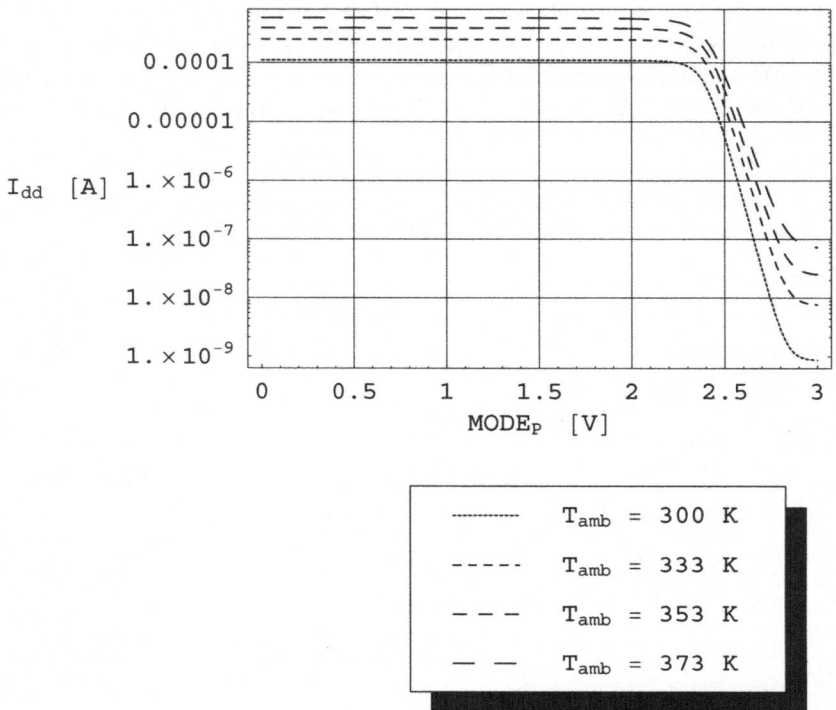

Figure 7.13.   Weak-inversion leakage current for 5000 parallel Triple-S latches as a function of the $MODE_P$ bias voltage for different ambient temperatures. Low and high threshold voltages are $200mV$ and $600mV$, respectively, and $U_{dd} = 2.5V$.

to the PMOS mode transistors, for different ambient temperatures T_{amb}. At the same time a voltage varying from $+2.5V$ to $-0.5V$ is applied to the $MODE_N$ bias pin (not indicated in figure 7.13). Hence, all mode transistors are gate biased. At room temperature ($300K$), when the latch is in the active period and idle, i.e. $MODE_P = 0V$, the leakage current is about $130\mu A$. In the standby period, i.e. $MODE_P = 3V$, the leakage is reduced to about $800pA$. This is a ratio of ample 5 decades, corresponding with the difference of $400mV$ between the high and low threshold voltages, assuming a sub-threshold slope of 80mV/decade. The numbers for the leakage currents can be verified with the presented measurement results for the gate biased minimum sized $0.25\mu m$ transistor of figure 7.11. The leakage of an idle latch in active periods is determined by the non-conducting low threshold voltage transistors, i.e. biased at $U_g = 0V$, of the "actual" inverters. Since 5000 latches contain 10000 inverters the leakage of one inverter will be around $13nA$. This is in accordance with figure 7.11, since for $U_g = 0V$ the drain current equals about $13nA$. The leakage current of the latch in standby periods is determined by the leakage of the parallel connected non-conducting high-threshold-voltage state switch and the low-threshold-voltage mode switch. The NMOS mode switch is biased at $U_g = -500mV$, whereas the PMOS mode switch is biased at $U_g = 3V$. The leakage of 5000 latches equals about $800pA$, hence the leakage of one inverter will be around $80fA$. This result cannot be deduced from figure 7.11, since the leakage current is limited by the measurement equipment. From figure 7.13 it can be seen that at a bias level of $MODE_P = 3V$ the leakage current is limited by the high threshold voltage transistor. As discussed in section 3.3.1.1 for higher temperatures the reduction factor reduces due to the increased weak-inversion slope and lowered threshold voltage, leading to exponentially increasing weak-inversion currents. Moreover, junction leakage currents also drastically increase with temperature eventually masking weak-inversion currents. However, as will be discussed in the next chapter, application of weak-inversion leakage current reduction techniques is only useful in systems possessing a large ratio between standby and active period times, providing for relatively low chip temperatures.

Figure 7.14 shows the supply current as a function of the PMOS source bias voltage, i.e. U_{dd}. The gate of the PMOS mode switch remains at $+2.5V$.

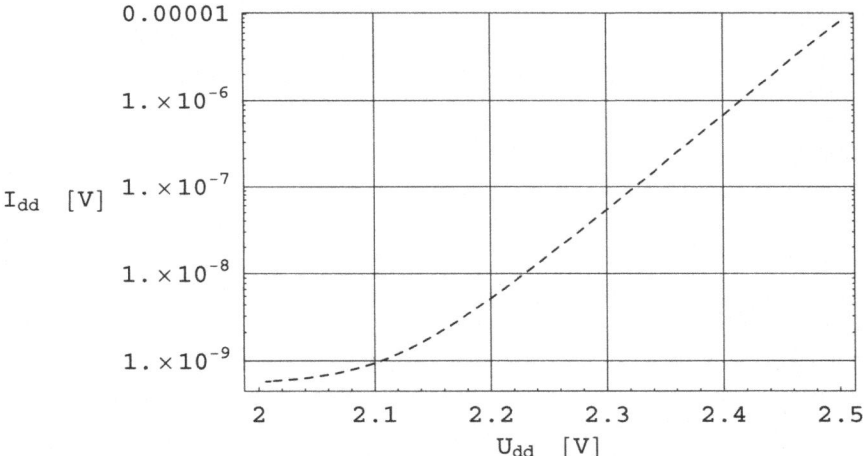

Figure 7.14. Leakage current for 5000 parallel Triple-S latches as a function of the U_{dd} bias voltage at $T_{amb} = 300K$. Low and high threshold voltages are $200mV$ and $600mV$, respectively.

Also a voltage varying from $0V$ to $+0.5V$ is applied to the sources of the NMOS transistors (not indicated in the figure), while the gate of the NMOS mode switch remains at $0V$. This procedure comes down to reduction of the voltage swing in standby periods, while the potentials of the $MODE_N$ and $MODE_P$ bias pins remain unchanged. Hence, all mode transistors are source biased. Weak-inversion currents are reduced from $130\mu A$, in idle active periods to 550pA in standby periods. Accordingly, the leakage current of one Triple-S inverter equals $55fA$. The leakage level is slightly lower than in the case of gate biasing, since GIDL currents of non-conducting transistors in series with the smart series switches, i.e. non-conducting transistors of the "actual" inverters, are reduced. Note from figure 7.13 that the leakage current already has been reduced from $130\mu A$ to $8\mu A$ when $MODE_P$ bias equals 2.5V, which equals the value for the leakage current for $U_{dd} = 2.5V$ in figure 7.14. From this figure it can be seen that at a bias level of $U_{dd} = 2V$ the leakage current is limited by the high threshold voltage transistor.

For ultra thin gate oxide transistors gate leakage currents are also reduced, since gate-source and drain-gate potentials are reduced. However, data integrity maybe compromised by the reduced voltage swing.

As can be seen from the measurement results shallow trench isolation currents scale proportionally with channel leakage currents and no sub-threshold humps, discussed in section 3.3.1.1 page 39, are present. Therefore, STI forms no obstacle in reducing weak-inversion currents using Triple-S.

7.3   Conclusions

In case a system's stored energy is not dissipated and no auxiliary circuitry is required to reduce weak-inversion leakage currents during standby periods, the energy saved in these periods only depends on their total time and is independent of the active/standby duty cycle. However, taking the energy stored in the system and the energy dissipated in the auxiliary circuitry into account, break-even standby times can be determined. Overhead costs, caused by the auxiliary circuitry, are dominant in practical cases.

Substrate biasing loses its effectiveness to reduce weak-inversion leakage currents, caused by down scaling of both applied voltages and the body factor.

Source biasing reduces weak-inversion currents exponentially, without introducing GIDL, by introducing a negative gate-source potential applied at the source terminal. Application of source biasing in power switches is straightforward, since their state is always the same during standby periods, i.e. non-conducting.

Gate biasing reduces weak-inversion currents exponentially by introducing a negative gate-source potential applied at the gate terminal. However, GIDL and gate leakage will increase for future technologies. Gate biasing is also applicable in power switches.

Triple-S is a very effective weak-inversion current reduction technique. Speed performance and leakage reduction can be designed independently from each other. A reduction of 5 decades in a latch has been demonstrated. Both source and gate biasing techniques are applicable for the mode switches.

8

TRIPLE-S CIRCUIT DESIGNS

In the previous chapter Triple-S has been introduced as a new weak-inversion leakage-current reduction technique. In this chapter Triple-S will be regarded from a practical point of view. In section 8.1 the implications of implementing Triple-S and power switches in a standard CMOS process flow will be discussed. Experimental circuits have been designed in an adapted 0.25μm technology to demonstrate among other things the effectiveness of Triple-S in practice. These circuits will be presented in section 8.2. According to the law of conservation of misery an improvement in leakage reduction will have to be exchanged for a deterioration of other parameters. First the effectiveness of leakage reduction by Triple-S will be considered. Thereafter, the penalty on speed, area and functional power dissipation is described in section 8.3. Finally, in section 8.4 the application of Triple-S in GSM and UMTS systems is evaluated.

8.1   Process flow

To verify the weak-inversion leakage-current reduction capabilities of Triple-S, experimental circuits have been processed in 0.25μm technology and tested. Here it will be discussed which process flow is necessary to implement Triple-S in digital CMOS circuits.

In active periods, depending on the state of the clock, system states are contained successively by master and slave latches of flip-flops. During standby periods system states are either retained by master or slave latches, all remaining circuit parts are switched off completely by power switches. Therefore, Triple-S only has to be applied either to master or slave latches of flip-flops. To implement this power reduction scheme the following is required:

- low- and high-threshold-voltage transistors;
- stable and defined clock signal during standby periods;

121

- "virtual" power supply lines;

- power switches.

Triple-S requires low- and high-threshold voltage transistors. Starting point was a standard single $570mV$ threshold voltage 0.25μm CMOS technology. For the experiments additional reticles were required to provide for the $600mV$ high threshold voltages. Moreover, the low threshold doping levels had to be adjusted to provide for a $200mV$ low threshold voltage, to make leakage levels comparable with those in 0.13μm technology and beyond. Multi thresholds are standard available from 0.18μm technology.

To keep the memory loop of a latch closed during standby periods, its clock inputs should be stable and well defined. Therefore, clock drivers are not allowed to be just disconnected from the power supply lines. Clock driver circuits will be discussed in more detail in section 8.2.

Circuits that are switched off during standby periods possess "virtual" power supply lines, which are connected via power switches to the actual power supply lines. Considering layout consequences when implementing "virtual" power supply lines two schemes are available:

- only flip-flops possess virtual power supply lines;

- all cells possess virtual power supply lines.

The consequence of the use of the scheme of figure 8.1, in which only flip-flops possess virtual power supply lines, is that always at least two rows of Triple-S flip-flops and two rows of arbitrary other standard cells need to abut. The reason is that the power line U_{dd} of the Triple-S flip-flops is not allowed to abut to the U_{ddv} line of all other standard cells. Advantages of this scheme are direct application of standard library cells lacking virtual power lines, saving design time creating low power libraries, and saving routing area. Disadvantages are the possible increase in average wire length, since placement of the cells becomes less flexible.

Application of virtual power lines to all standard cells allows very flexible cell placement and routing, since cell placement becomes independent of cell abutment, indicated in figure 8.2. However, this flexibility is gained at the cost of extra design effort creating new low-power library cells, reduced routing space and increased cell area. The latter is caused by the need for separate connections of sources to U_{ddv} lines and of substrates to U_{dd} lines. This scheme has been applied to all experimental designs, discussed in the next section, to make as much use of standard place and route tools as possible.

Power switches can be implemented with either NMOS or PMOS transistors. The choice is arbitrary. Although the conductivity of PMOS transistors is somewhat smaller compared to NMOS transistors, caused by the lower mobility

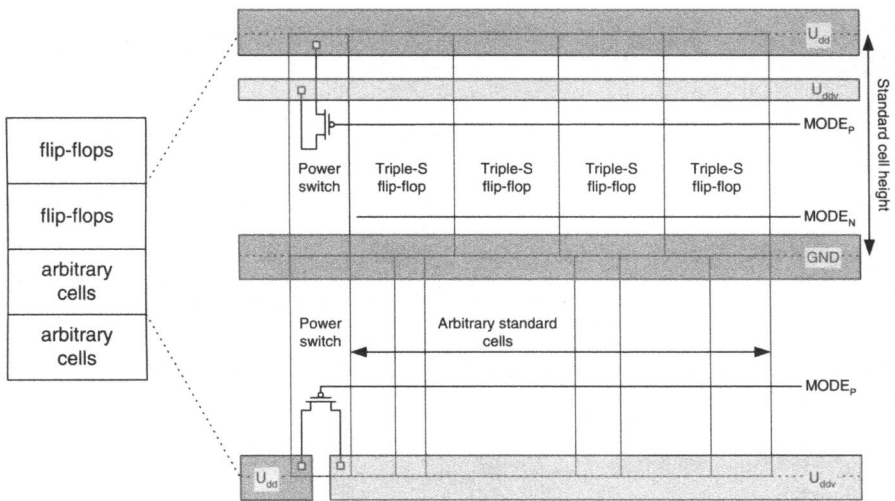

Figure 8.1. Virtual power line scheme applied to Triple-S flip-flops only.

of holes, a larger part of the cell height is reserved for PMOS allowing them to become wider for equal conductivity. Application of both NMOS and PMOS power switches is not useful, since equal weak-inversion current levels can be obtained for less area overhead with one-sided power switches.

8.2 Experimental circuits

Experimental circuits have been designed in 0.25μm standard single adapted low threshold voltage, i.e. $U_{th} = 200mV$ and dual threshold voltage CMOS technology. The high-threshold voltage equals 600mV and low-threshold voltage equals 200mV to show compatibility of the Triple-S technique with future process technologies. The following experimental circuits have been designed and processed:

- latches (leakage);
- flip-flops (leakage, area);
- shift registers (leakage, area);
- ring oscillators (speed).

Figure 8.3 depicts a standard positive edge triggered flip-flop, whereas figure 8.4 shows a positive edge triggered Triple-S flip-flop. The slave latch has been implemented with Triple-S inverters. All low threshold-voltage transistors in all latches have a W_{ch}/L_{ch} ratio of $0.35\mu m/0.25\mu m$, whereas the high-threshold voltage transistors have a W_{ch}/L_{ch} ratio of $0.35\mu m/0.35\mu m$. All NMOS

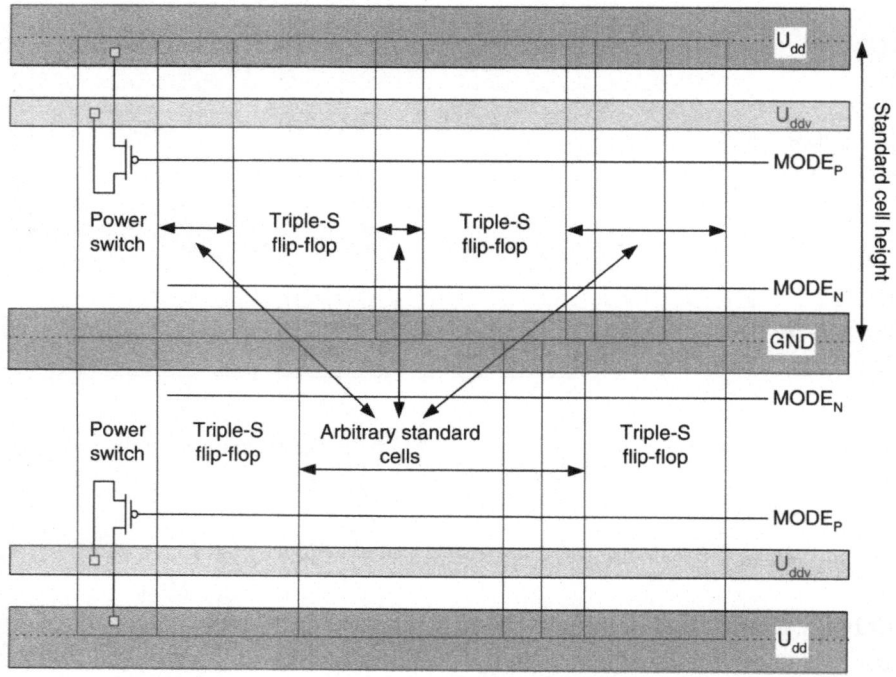

Figure 8.2. Virtual power line scheme applied to Triple-S flip-flops mixed with other standard cells, for easy abutment.

substrate contacts are connected to GND and all PMOS substrate contacts are connected to U_{dd}. For PMOS transistors with their sources connected to U_{ddv} the substrate contacts are connected to U_{dd}, otherwise the substrate short circuits the power switches. All flip-flops are implemented with a multiplexed data (D) and test (TI) input, to enable the formation of scan chains, as discussed in section 6.1.1.1. The test enable (TE) signal selects whether input D or TI is latched. A clock driver is, like the test enable inverter, integrated in each flip-flop standard cell. This clock driver is implemented in such a way that a stable low clock signal CK during standby periods ($\overline{CP}$ is high and CP is low) keeps the loop of the slave latch closed while its weak-inversion currents are reduced by both the switch controlled by $MODE_N$ and the global PMOS power switch.

With this flip-flop 512-bits shift registers and 16 parallel connected 32-bits binary counters have been made. These numbers of cells have been chosen to enable leakage current measurements, i.e. currents should be well above $1pA$.

Ring oscillators have been designed to determine power-delay products and compare speed performance between standard inverters and Triple-S inverters. Each ring oscillator consists of 4001 sections to provide for an oscillation frequency between 500 kHz and 3 MHz for high- and low-threshold-voltage

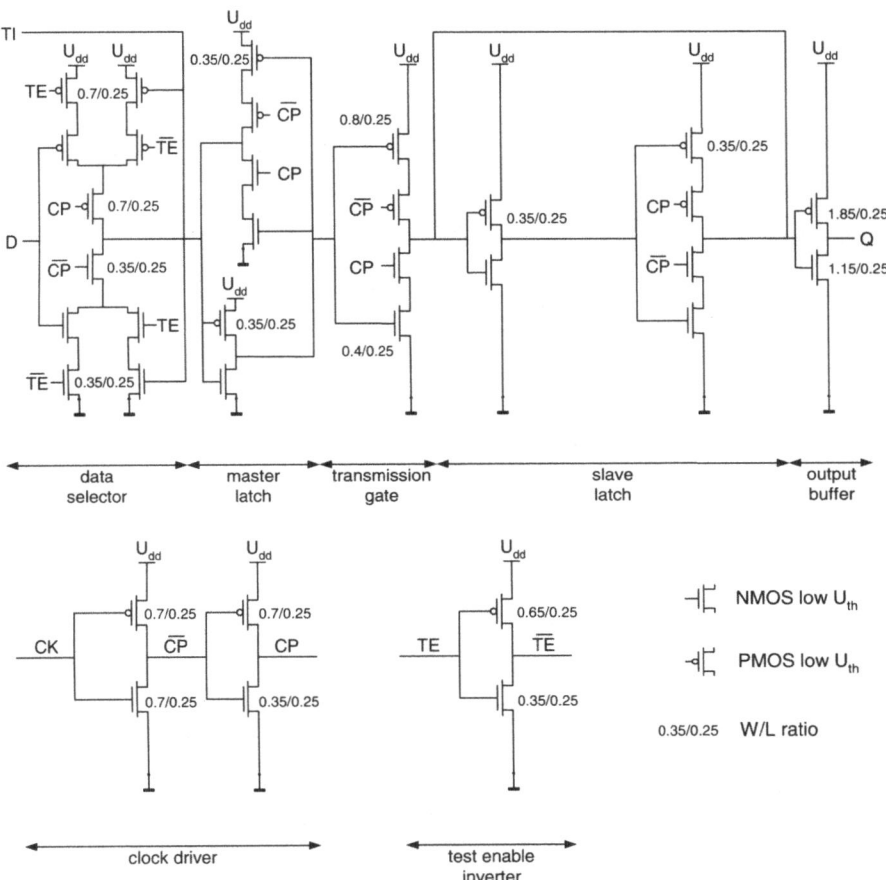

Figure 8.3. Positive edge triggered standard flip-flop.

transistors, respectively for a supply voltage of $2.5V$. The ring oscillators will be discussed in the next section.

8.3 Leakage, speed, area and functional power

In section 8.3.1 through 8.3.4 the impact of the Triple-S technique on leakage, speed, area and functional power [69] will be discussed, respectively. This power reduction scheme comprises that all combinatorial logic can be switched off completely during standby periods, while Triple-S latches provide for leakage reduction and data retention. However, the extra transistors needed to implement this power reduction scheme reduce speed performance, increase chip area and functional power dissipation.

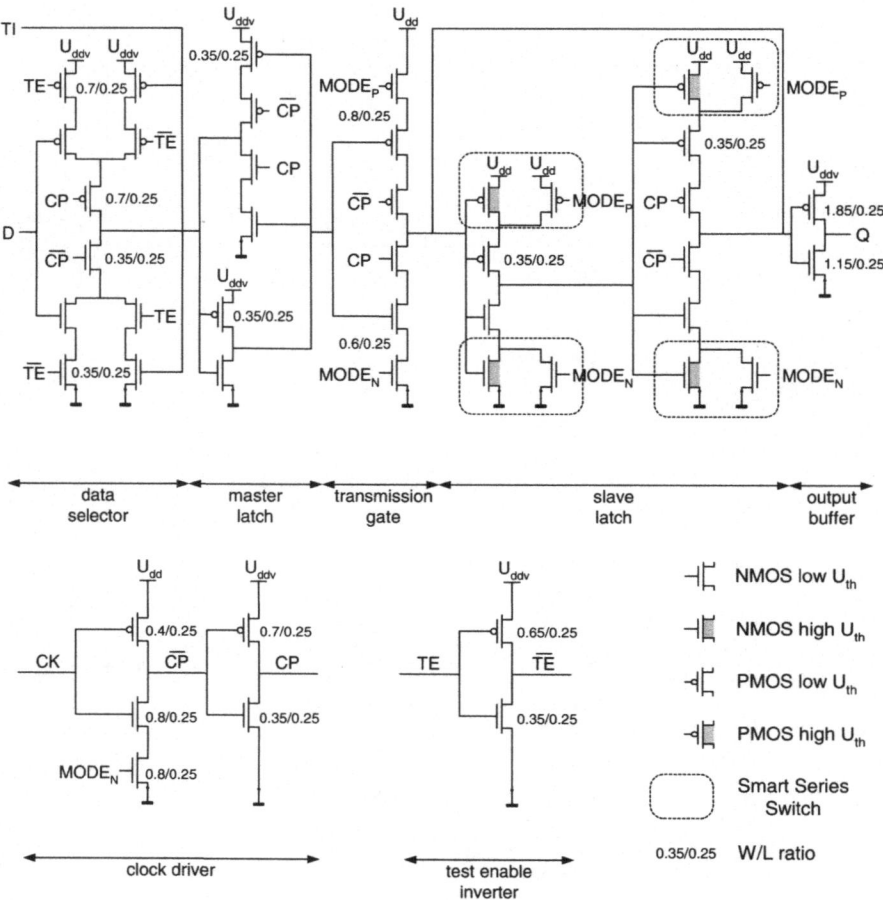

Figure 8.4. Positive edge triggered Triple-S flip-flop.

8.3.1 Leakage

The performance of weak-inversion leakage current reduction for Triple-S latches has been presented in section 7.2.3. Figure 8.5 presents leakage current reduction of a 512 bits Triple-S shift register under gate biasing conditions of the mode switch at $T_{amb} = 300K$. The indicated leakage current represents the measured supply current, I_{dd}, as a function of the $MODE_P$ bias voltage, which is applied to the PMOS mode transistors. At the same time a voltage varying from $+2.5V$ to $-1.5V$ is applied to the $MODE_N$ bias pin (not indicated in figure 8.5). Hence, all mode transistors are gate biased. For a $MODE_P$ bias voltage of 3V, the weak-inversion leakage current of the shift registers has been reduced from $20mA$ to $20\mu A$, which is a factor of 1000.

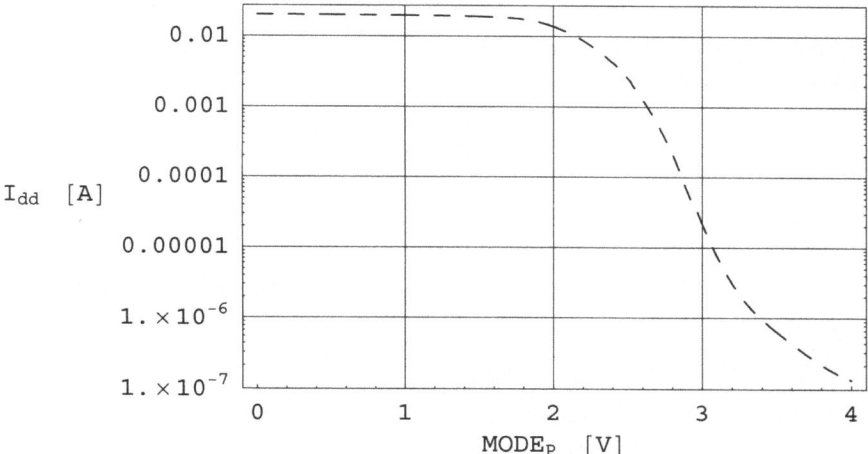

Figure 8.5. Weak-inversion leakage current for a 512 bits Triple-S shift register as a function of the $MODE_P$ gate bias voltage at $T_{amb} = 300K$. Low and high threshold voltages are $200mV$ and $600mV$, respectively.

Figure 8.6 indicates the leakage current reduction of a 512 bits Triple-S shift register in case of source biasing of the mode switch. The indicated leakage current represents the supply current, I_{dd}, as a function of the PMOS source bias voltage, i.e. U_{dd}. The gate of the PMOS mode switch remains at $+2.5V$.

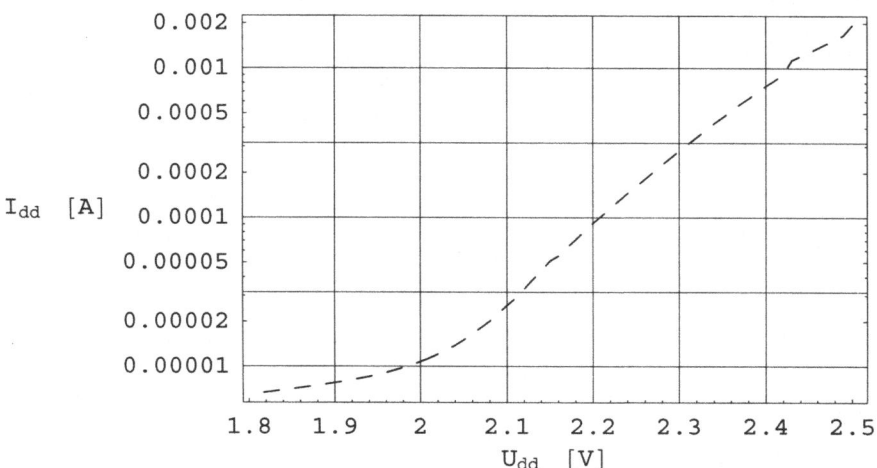

Figure 8.6. Weak-inversion leakage current for a 512 bits Triple-S shift register as a function of the, U_{dd}, source bias voltage. Low and high threshold voltages are $200mV$ and $600mV$, respectively.

Also a voltage varying from $0V$ to $+0.7V$ is applied to the sources, i.e. U_{GND}

of the NMOS transistors (not indicated in the figure 8.6), while the gate of the NMOS mode switch remains at $0V$. At 2.5V the mode transistors are already switched off, which reduces leakage of the Triple-S shift registers from about $20mA$ to $2mA$, as can be seen from figure 8.5. For a swing of only $0.5V$ total leakage current reduction is about a factor of 2000, i.e. from $20mA$ in idle active periods to $10\mu A$ at $U_{dd} = 2V$. Data is still retained at values for $U_{dd} = 1.8V$ and $U_{GND} = 0.7V$.

Triple-S reduces weak-inversion leakage in inverters and latches. It should be noted that leakage paths might still exist even when Triple-S has been applied. In general it can be stated that *where drains meet leakage paths exist*. This is already true for a simple CMOS inverter. Addition of Triple-S solves the problem within the inverter. However, more tricky and unexpected leakage paths occur in flip-flops, in which data dependencies also play an important role. The connected drains forming the output of the Triple-S latch are connected to the drains forming the output of the clocked inverter preceding this latch. Besides application of Triple-S weak-inversion leakage currents can flow between the power supply lines, via the path indicated by the dotted lines with arrow heads in figure 8.7, when transistors T_P and T_N are not added to the clocked inverter preceding the Triple-S latch. Accordingly, both smart series switches and transistors T_P and T_N reduce the weak-inversion leakage currents in latches and transmission gates, respectively. The data dependent leakage

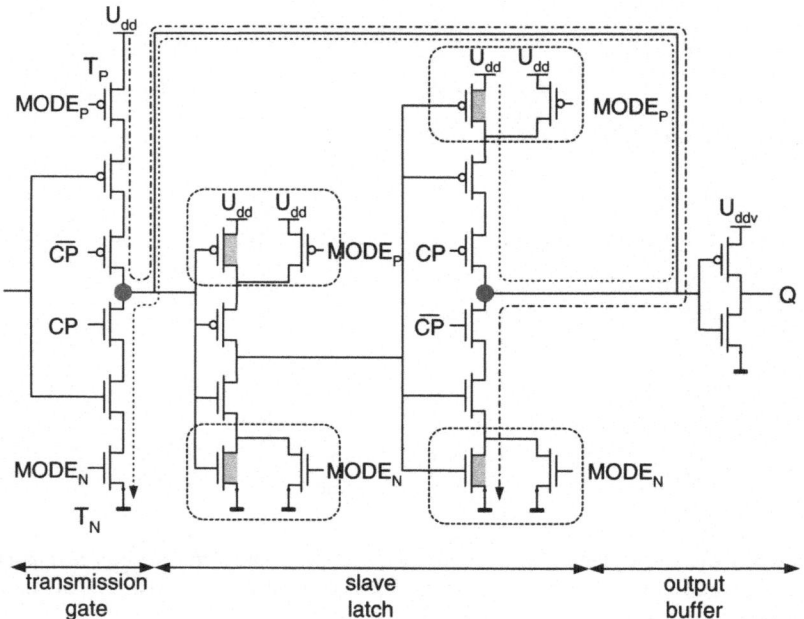

Figure 8.7. Paths of reduced weak-inversion currents in the Triple-S latch connected to a transmission gate.

path is reduced by T_P. Transistor T_P has to be added to the transmission gate to prevent the latch from charging the virtual power supply line U_{ddv}, in case the logic state of output Q is high. Leaving out T_P would introduce the leakage path indicated in figure 8.8. For his reason transistor T_P is not allowed to be

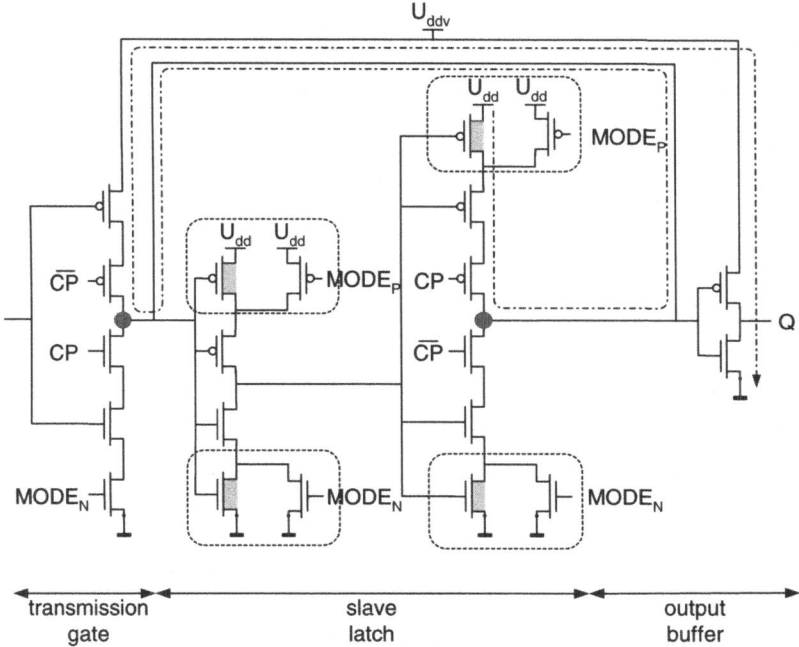

Figure 8.8. Virtual power supply line is charged by the latch when the logic state of output Q is high.

combined with the general power switch.

8.3.2 Speed

The propagation delay of a logic gate is a measure for its speed performance. The 0.25μm-technology ring oscillators, depicted in figure 8.9, have been used to determine propagation delays of standard and Triple-S inverters. The standard inverter structures have been implemented with both low- and high-threshold-voltage transistors. Each ring oscillator contains 4000 inverter sections, resulting in oscillation frequencies, at a nominal supply voltage of $2.5V$, of $780kHz$, $960kHz$ and $2.3MHz$ for the high threshold voltage, Triple-S and low-threshold voltage inverters, respectively. Figure 8.10 depicts their propagation delays as a function of the power supply voltage. From this figure it can be seen that for $U_{dd} = 2.5V$ the propagation delay of the Triple-S inverter is about 2.4 times larger than the propagation delay of the low-threshold standard inverter. However, it is about 1.2 times smaller than the propagation delay of

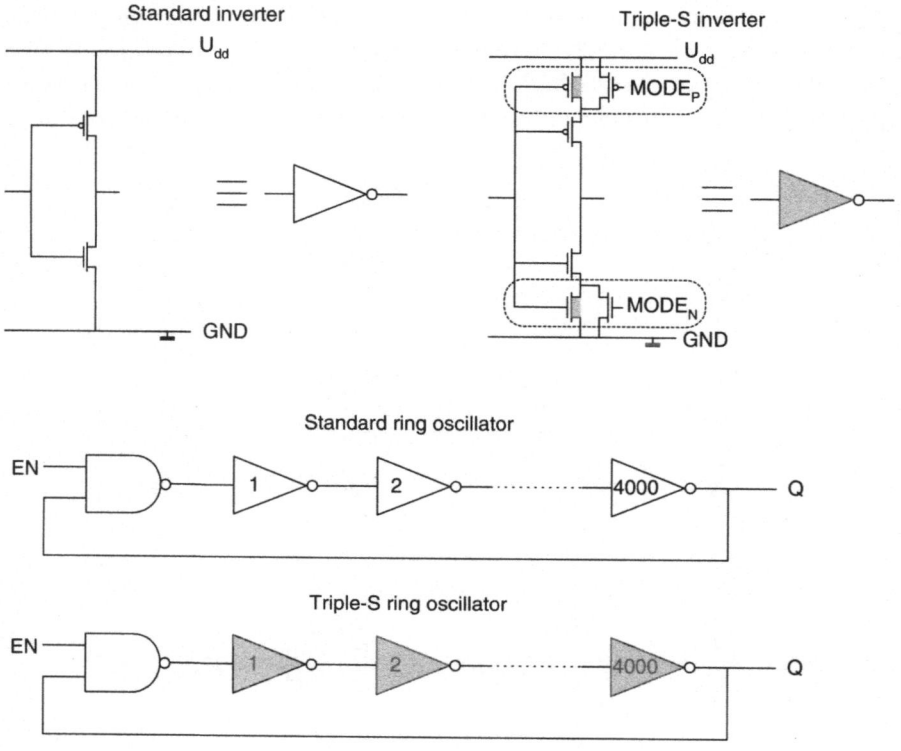

Figure 8.9. Standard and Triple-S ring oscillators.

the high-threshold standard inverter. Compared to the standard inverters, for lower supply voltages the difference in propagation delay remains almost constant for Triple-S inverters, whereas for high threshold inverters delays increase strongly. This aspect favors for Triple-S instead of the use of high thresholds only. Since Triple-S latches still put quite a heavy burden on propagation delays, they should be left out of critical paths in flip-flops, as is the case in the experimental flip-flop introduced in figure 8.4. Table 8.1 indicates the speed penalty for a Triple-S flip-flop compared to the "standard" 0.25μm-technology flip-flop. The schematic of the positive edge triggered standard flip-flop is depicted in figure 8.3. The propagation delay of the flip-flop is defined here as the

Flip-flop	Propagation delay [ps]
Standard	750
Triple-S	825

Table 8.1. Speed penalty for the Triple-S flip-flop.

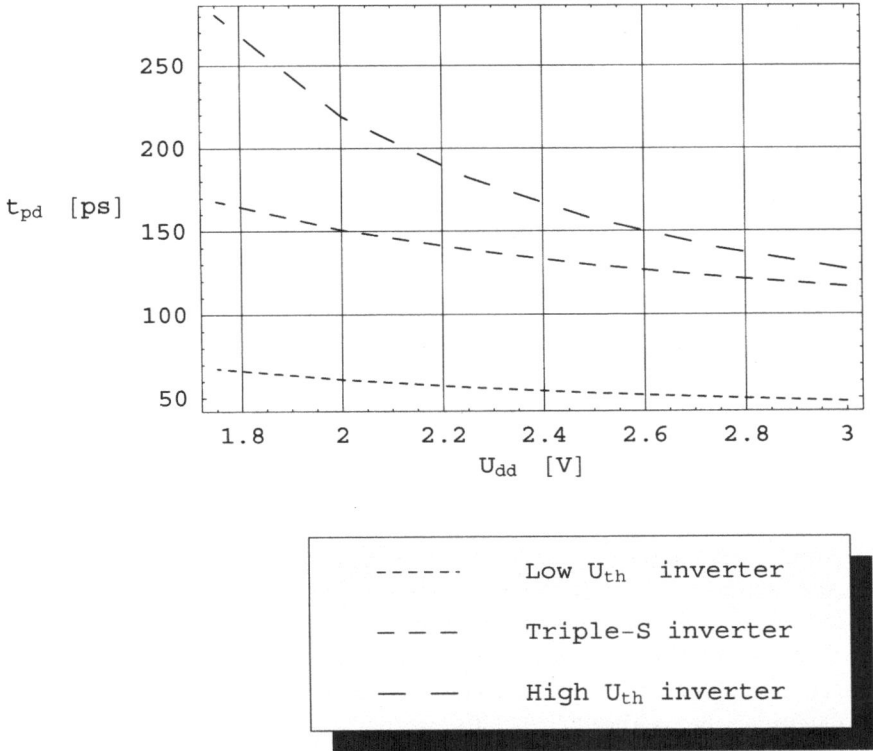

Figure 8.10. Propagation delays as a function of power supply voltage for standard low- and high U_{th} inverters and Triple-S inverters. The nominal supply voltage equals $2.5V$.

elapsed time between the rising edge of the clock signal at input CK and the change of the data at output Q, see figures 8.3 and 8.4. The presented propagation delays have been determined by calculation and simulation using results gained from both ring oscillator and transistor characterization measurements. The fanout of the flip-flops has been assumed to be equal to 10 standard inverters. Under this condition the propagation delay of the Triple-S flip-flop is about 10% larger compared to the standard flip-flop. The impact of this extra delay on total performance depends on the number of logic gates between flip-flops, i.e. the logic depth. Decreasing the logic depth boosts performance, but increases the number of flip-flops and therefore also power dissipation.

8.3.3 Area

Application of the Triple-S technique brings extra occupation of chip area with it. Table 8.2 shows the area occupied by a single flip-flop, a standard shift

	Standard [μm^2]	Triple-S [$(\mu m)^2$]	Extra area [%]
flip-flop	144	198	38
32 bit shift register	4644	6480	40
32 bit binary counter	9126	10962	20

Table 8.2. Area occupied by a single flip-flop, a shift register and a binary counter implemented with and without Triple-S. For the shift register and binary counter the area of the power switches is included.

register and a binary counter and their Triple-S versions. The presented areas include power switches. However, in general on average about 30% of the chip area is occupied by flip-flops. Therefore, the extra amount of area occupied by implementing Triple-S flip-flops and power switches equals about 12%.

8.3.4   Functional power

Addition of smart series switches to flip-flops will introduce overhead costs, i.e. the functional power dissipation increases. In order to determine the difference in functional power dissipation between standard and Triple-S flip-flops, the functional power dissipation of shift registers of both types will be determined. The procedure is discussed first, thereafter measurement results are used to determine the overhead costs. To distinguish the contribution to the functional power dissipation of the clock circuitry and the clock bond pad driver, measurements on different numbers of standard flip-flops have been performed under zero data activity. This produces the switched capacitances for clock circuitry and a bond pad driver. Thereafter, maximum data activity measurements yield the total switched capacitance of the data circuitry. Finally, zero and maximum data activity measurements on Triple-S flip-flops yield total switched capacitances for clock and data circuitry. Now the extra contribution of Triple-S switches to the functional power dissipation can be calculated.

Measurement results can be used to determine the overhead costs. Figure 8.11 shows the measurement results of the functional power dissipation as a function of clock frequency, f_{clk}, for both 512 and 2048 bits shift registers using standard flip-flops. The power supply voltage equals 2.5V. From the 512 and 2048 bits zero data activity curves the capacitance of the clock circuitry for a 512 bits shift register, $C_{clk\_std,tot}$, and the bond pad driver of the clock signal, C_{pad}, can be determined. For the functional power dissipation of the 512 bits standard shift register under zero data activity conditions holds:

$$P_{func\_std,0} = f_{clk} \left(C_{clk\_std,tot} + C_{pad} \right) U_{dd}^2 \qquad (8.1)$$

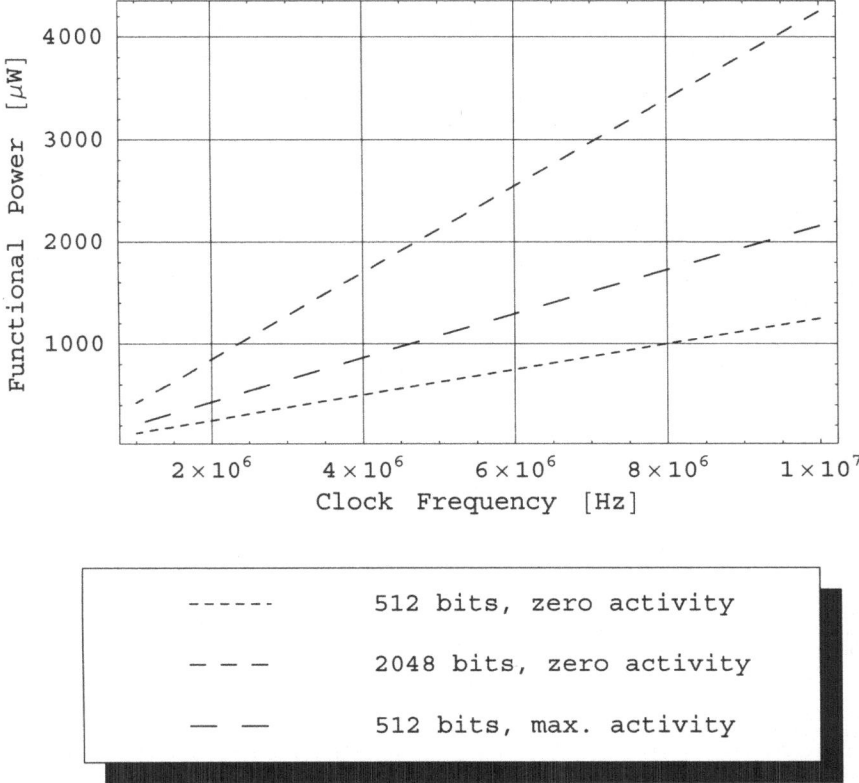

Figure 8.11. Functional power dissipation of a 512 and 2048 bits shift register consisting of standard flip-flops for cases of zero and maximum data activity. The supply voltage equals 2.5V.

The functional power dissipation of the 2048 bits standard shift register under zero data activity conditions can be expressed as:

$$P_{func\_std\_2048bits,0} = f_{clk}\left(4C_{clk\_std,tot} + C_{pad}\right)U_{dd}^2 \qquad (8.2)$$

From equation 8.1 and 8.2 and figure 8.11 both $C_{clk\_std,tot}$ and C_{pad} can be determined and equal approximately $17pF$ and $4pF$, respectively. The functional power dissipation of a 512 bits standard shift register under maximum data activity, i.e. the data changes every clock cycle, can be expressed as:

$$P_{func\_std,max} = f_{clk}\left(\underbrace{C_{clk\_std,tot} + C_{pad}}_{clock\ circuit} + \frac{1}{2}\left(\underbrace{C_{data\_std,tot} + 2C_{pad}}_{data\ circuit}\right)\right)U_{dd}^2$$

$$(8.3)$$

in which $C_{data\_std,tot}$ represents the total switched capacitance when all data bits change their state every clock cycle. Accordingly, under maximum data activity conditions, its switching frequency equals half the clock frequency. The total pad capacitance comprises the data input and output pads, switching at half the clock frequency, and the clock pad switching at the clock frequency. The total switched data capacitance for a 512 bits standard shift register equals about 21pF, which can be determined from equation 8.3, figure 8.11 and the values already determined for $C_{clk\_std,tot}$ and C_{pad}.

Figure 8.12 shows the functional power dissipation as a function of clock frequency, f_{clk}, for a 512 bits shift register using Triple-S flip-flops. The power

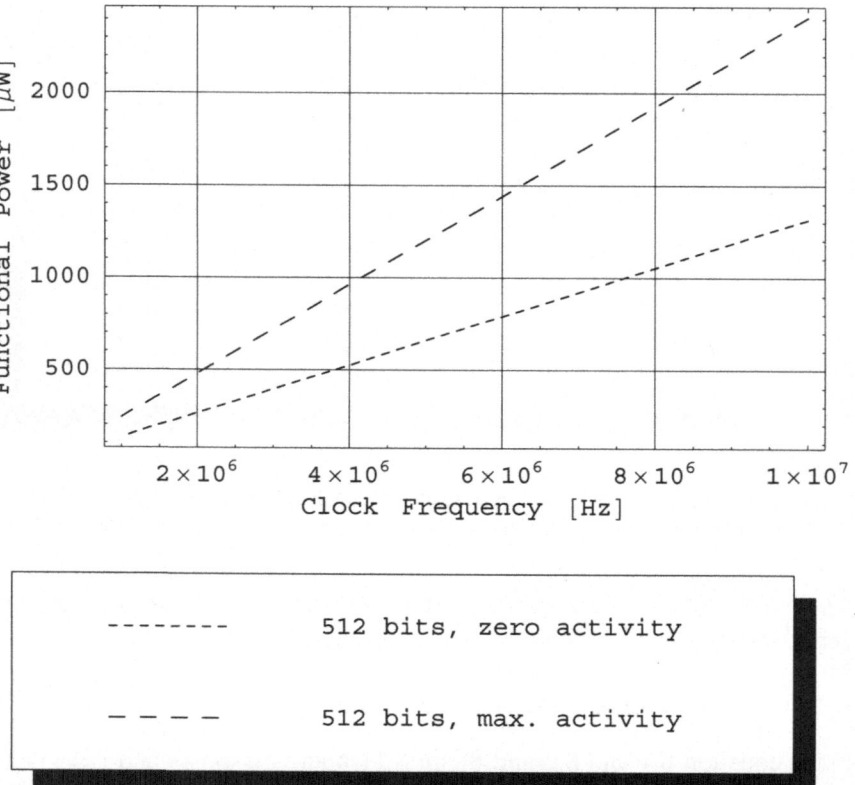

Figure 8.12. Functional power dissipation of a 512 bits shift register consisting of Triple-S flip-flops for cases of zero and maximum data activity. The supply voltage equals 2.5V.

supply voltage equals 2.5V. The functional power dissipation of the 512 bits Triple-S shift register under zero data activity conditions can be expressed as:

$$P_{func\_TS,0} = f_{clk} \left(C_{clk\_TS,tot} + C_{pad} \right) U_{dd}^2 \tag{8.4}$$

The switched clock capacitance of the Triple-S shift register $C_{clk\_TS,tot}$ equals approximately 17pF, which can be determined, as in the previous case, from equation 8.4, the previously determined value for C_{pad} and figure 8.12. For the functional power dissipation of a 512 bits Triple-S shift register under maximum data activity holds:

$$P_{func\_TS,max} = f_{clk} \left(\underbrace{C_{clk\_TS,tot} + C_{pad}}_{clock\ circuit} + \frac{1}{2} \left(\underbrace{C_{data\_TS,tot} + 2C_{pad}}_{data\ circuit} \right) \right) U_{dd}^2$$

(8.5)

in which $C_{data\_TS,tot}$ represents the total switched capacitance when all data bits change their state every clock cycle. The total switched data capacitance for a 512 bits Triple-S shift register equals about 27pF, which can be determined from equation 8.5, figure 8.12 and the values already determined for $C_{clk\_TS,tot}$ and C_{pad}.

From the values determined for the switched clock and data capacitances the functional power dissipation, under maximum data activity condition, and without bond pads can be calculated for the 512 bits standard and Triple-S shift registers according to:

$$P_{func,max} = f_{clk} \left(C_{clk,tot} + \frac{1}{2} C_{data,tot} \right) U_{dd}^2$$

(8.6)

in which $C_{clk,tot}$ and $C_{data,tot}$ represent the switched clock and data capacitance for either standard or Triple-S shift register, respectively. The resulting plot is indicated in figure 8.13. From this plot it can be seen that the overhead costs, under maximum data activity conditions, i.e. worst case, for Triple-S shift registers is about 14%. Since a shift register is nothing more than a chain of flip-flops, the same conclusion holds for Triple-S flip-flops. On average 30% of the chip area is occupied by flip-flops resulting in about 4% overhead costs.

8.4 Practical applications and limitations

The break-even standby time of the Triple-S technique will be considered for $0.25\mu m$, $0.18\mu m$, and $0.13\mu m$ CMOS technologies in both GSM and UMTS applications [79]. Table 8.3 shows a summary of the results.

Both applications have active modes and low-power modes. In the active mode all parts of the system are fully activated and the system is able to send and receive information. The low-power mode consists of an active period (T_{active}, see figure 2.3 for definitions), needed for paging related activities, and a standby period ($T_{standby}$) where the whole system is at rest, i.e. only the necessary system states are retained by flip-flops and the combinatorial logic is switched off. The increase of the logic gate capacitance in a Triple-S flip-flop equals about 14%, as has been discussed in the previous section. On average

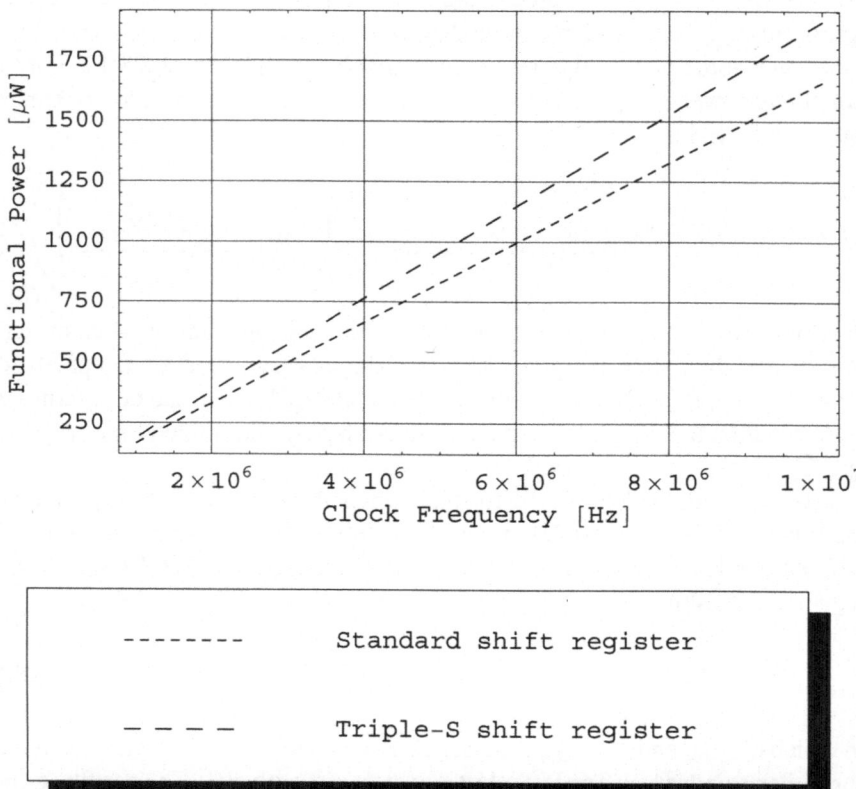

Figure 8.13. Functional power dissipation of a 512 bits shift register consisting of standard and Triple-S flip-flops for maximum data activity without bond pads. The supply voltage equals 2.5V.

Generation [μm]	0.25	0.18	0.13
$W_{ch}[\mu m]$	0.35	0.25	0.18
$U_{sw}[V]$	2.5	1.8	1.5
$C_g[fF]$	4	2	1.5
$I_{wi,U_{th,low}}[pA]$	2.5	100	1300
$GSM : T_{stdb,be\_oh}[s]$	1920	17	0.8
$UMTS : T_{stdb,be\_oh}[s]$	64	0.58	0.027

Table 8.3. Minimum break-even standby times for the Triple-S technique, with $\delta = 0.04$, $\alpha = 1$ and $f_{clk} = 200MHz$.

flip-flops occupy approximately 30% of the chip area. Consequently, the total relative increase in the logic gate capacitance equals 4%, causing an equal increase in the functional power dissipation, i.e. overhead costs ($\delta = 0.04$), as has been elaborated on in section 7.1.3. In the standby mode of a mobile GSM station the active period equals 6 Time Division Multiple Access (TDMA) frames and the standby period equals N*45 TDMA frames. The length of one TDMA frame is approximately $5ms$ and N ranges between 2 and 6. Thus, in the standby mode the active period equals $30ms$ and the standby period equals at least $450ms$. In UMTS the active period during standby mode is not fixed but can be between about $100\mu s$ and $10ms$. Realistic active and standby times are for instance $1ms$ and $700ms$, respectively. Table 8.3 shows the minimum break-even standby times for these systems realized in several technologies. As discussed in section 7.1.3 the minimum break-even standby time can be determined as follows (see equation 7.9):

$$\frac{\overline{T}_{stdb,be\_oh}}{\overline{T}_{stdb,be\_se}} = 2\delta f_{clk,eq}\overline{T}_{active} \qquad (8.7)$$

in which $\overline{\alpha}$ is assumed one and the voltage swing is assumed to be constant. A system clock frequency $f_{clk,eq}$ of $200MHz$ has been assumed for both applications in all technologies. From table 8.3 it can be seen that the reduction technique is not effective in GSM applications, mainly because of the relatively long active periods in standby mode. For UMTS applications, however, in 0.13μm technology it is quite effective. In the latter case the 4% overhead costs are compensated in $27ms$ (T_{stdb,be_oh}). In the remainder of the standby time, i.e. $25 * T_{stdb,be\_oh}$, an energy amount equal to the functional energy dissipated in the active period of the standby mode is saved ($25 * 4\% = 100\%$). Accordingly the standby time can be doubled.

From this example it can be deduced that application of the Triple-S technique is only useful in systems possessing a large ratio between the standby and active period times. A collateral advantage of a large ratio between standby and active period times is that chip temperatures remain low, reducing the absolute weak-inversion current level and the weak-inversion slope, providing for the largest reduction factor as has been shown in figure 7.13.

8.5 Conclusions

Implementation of Triple-S in combination with power switches requires adaptations of the standard CMOS process flow. The following is required:

- low- and high-threshold-voltage transistors;

- stable and defined clock signal during standby periods;

- "virtual" power supply lines and power switches.

Experimental circuits have been processed in 0.25μm dual threshold technology. The propagation delay of a Triple-S flip-flop is 10% larger than its "standard" counter part. Circuits containing Triple-S flip-flops and power switches on average require 12% more chip area. Their overhead costs, i.e. extra functional power dissipation, total up to 4%, while their weak-inversion leakage current can be reduced by a factor of 1000.

Application of Triple-S is only useful in cases where standby periods are longer than the break-even standby time in case of overhead costs. Only then, the energy saved during standby periods makes up for the extra energy consumed during active periods, i.e. the overhead costs. It has been argued that Triple-S could be applied in UMTS applications, since their ratios between active and standby periods provide for favorable energy saving results.

9

CONCLUSIONS

For present and future CMOS technologies power consumption and power densities increase tremendously, caused by decreasing feature sizes and increasing clock frequencies, which on their turn are pushed by market demands for ever increasing chip functionality. Increased power consumption is responsible for increased operational costs, burdens the environment and shortens battery lifetime. Increased power densities make heat transfer from the chip a real challenge.

The majority of present-day computational systems consist of irreversible logic CMOS blocks. Hence, all power delivered to them will eventually be dissipated, either in the form of functional power or in parasitical power.

The functional power dissipation can be reduced by reducing the node transition-cycle activity factor, the clock frequency and the transition-cycle energy. Reducing the supply voltage or voltage swing is the most effective way of reducing the transition-cycle energy, because of its quadratic dependency.

The parasitical power dissipation is dominated by weak-inversion and gate leakage currents. Both increase exponentially due to scaling trends.

Scaling reduces device dimensions, the gate oxide included. For reliability reasons supply voltages have to be reduced. In order to maintain speed performance, threshold voltages need to be scaled down. Accordingly, scaling trends result in exponentially increasing weak-inversion and gate leakage currents. This is called the speed-performance leakage-power conflict. Weak-inversion and gate leakage currents become dominant in standby periods, since the functional and the short-circuit power dissipation are zero then. Therefore, weak-inversion and gate leakage currents will become a significant burden for especially battery operated devices possessing standby modes. Solutions are required to reduce these leakage currents in order to enlarge battery lifetime of mobile devices.

139

Regarding technological developments it seems that gate leakage will most probably be solved by the application of high-permittivity gate insulators.

During standby periods weak-inversion currents can be reduced in logic circuits with and without state retention. Reduction without state retention is performed by power switches and applied to combinatorial circuitry. For reduction with state retention a few options are available. System states can be stored in intrinsically low leakage memory like e.g. flip-flops or memory cells consisting of high-threshold-voltage transistors. Another option is to reduce the leakage of the individual transistors in cells containing the system states. A few possibilities are available to reach this goal. A common technique to reduce weak-inversion leakage currents is substrate biasing. However, the body factor of future CMOS technologies diminishes, resulting finally in negligible leakage current reduction. Regarding source and gate biasing techniques in sequential circuitry, no practical solutions have been found yet to implement these sources efficiently. However, both techniques can be applied in power switches and Triple-S mode switches. Triple-S, a new leakage current reduction technique, enables almost independent design of the leakage reduction factor and speed performance. To implement this power reduction scheme in combination with power switches, the following is required:

- low- and high-threshold-voltage transistors;

- stable and defined clock signal during standby periods;

- "virtual" power supply lines and power switches.

Leakage reduction factors between 10^3 and 10^5 have been demonstrated for circuits containing flip-flops and latches. These experimental circuits have been processed in an adapted $0.25\mu m$ CMOS technology with low and high threshold voltages of 200mV and 600mV, respectively. Compared to "standard" designs, implementation of Triple-S and power switches requires on average 12% more chip area and consumes 4% more functional power, i.e. overhead costs. Propagation delays of Triple-S flip-flops increase by 10%. Overhead costs are dominant compared to the energy stored in the system. To regain the overhead costs, weak-inversion currents have to be reduced considerably and standby periods have to be large compared to active periods.

10

SUMMARY

In 1965 Gordon Moore predicted that the total number of devices on a chip would double every year until the 1970s and every two years in the 1980s. This prediction is widely known as "Moore's Law" and eventually culminated in the Semiconductor Industry Association (SIA) technology road map. The SIA road map has been a guide for the industry leading them to continued wafer and die size growth, increased transistor density and operating frequencies, and defect density reduction. Semiconductor industry tries to keep up with Moore's law, to appease the hunger for ever increasing chip functionality. Accordingly, as discussed in Chapter 2, past, present and future CMOS technology generations are characterized by decreasing feature sizes, increasing clock frequencies and transistor densities causing both increased power consumption and power density. Especially for battery operated systems increased power consumption burdens the environment and is also responsible for increased operational costs, whereas increased power densities push chip temperatures to physical limits. Battery operated systems and nowadays desktop machines possess standby modes to reduce overall power consumption.

The power dissipation of computational processes depends on the physical processes on which they are based. They are either reversible or irreversible, as discussed in Chapter 3. In contrast to an irreversible process, a reversible process can traverse all states in opposite direction in such a way that it dissipates no power and does not increase the entropy of the universe when returning to its initial state. In case logical operations performed by reversible logic are physically being performed quasi-statically, the process will be physically reversible. However, practical operations will have to be performed in a limited amount of time, i.e. non quasi-statically, and hence become physically irreversible.

In literature the concept adiabatic often is used as a synonym for loss-less. However, adiabatic processes are not necessarily physically reversible and therefore not necessarily lossless.

Irreversible logic is based on physically irreversible processes, since after the logic operations the logic state sequences are not traversed in reverse order to return to initial states. Accordingly, logical irreversibility implies physical irreversibility. Nowadays the majority of computational systems consists of irreversible logic CMOS blocks. Consequently, all power delivered to them will eventually be dissipated, either in the form of functional power or in parasitical power.

The functional power is dissipated to just attribute to state changes of a digital CMOS circuit in favor of logic operations, and is a function of the following parameters:

- node transition-cycle activity;

- clock frequency;

- transition-cycle energy.

In Chapter 4 it has been discussed that the node transition-cycle activity can be reduced by e.g. choice of algorithm and reduction of signal skews. Parallelisation is a method to reduce the clock frequency. Reducing the supply voltage or voltage swing, statically or as a function of e.g. workload, is the most effective way of reducing the transition-cycle energy, because of its quadratic dependency. Increasing the reversibility factor by the use of ramp-wise, step-wise or resonant charging circuits, is another solution to reduce transition-cycle energy.

The parasitical power is either dissipated when the circuit is idle, defined as leakage power, or could be dissipated during state transitions without attributing to the actual changes of the internal states, defined as short-circuit power. Accordingly, the parasitical power is divided into two groups:

- leakage power dissipation;

- short-circuit power dissipation.

Leakage power is device related power, whereas short-circuit power is circuit related power caused by rail-to-rail currents during state transitions of static CMOS circuits. The latter can be reduced by preventing direct paths from rail to rail, e.g. by using domino logic or reducing the supply voltage below the sum of both NMOS and PMOS threshold voltages. The leakage power dissipation becomes dominant during idle periods, since the functional and short circuit power dissipation are zero then. It has been subdivided into three sub-groups distinguished by their origins:

- channel leakage current;

- diode leakage current;

- gate leakage current.

In Chapter 5 techniques have been presented to reduce the parasitical power dissipation. It has been argued that the weak-inversion current component of the channel leakage current and the gate leakage currents become dominant compared to the other components, since both increase exponentially due to device scaling trends.

Regarding technological developments it is to be expected that gate leakage will be reduced by application of high-permittivity gate insulators.

Chapter 6 introduced a classification of weak-inversion current-reduction techniques. These techniques might all be applied to systems possessing standby periods. At the highest level a distinction has been made between two main weak-inversion current-reduction concepts:

- power reduction without state retention;

- power reduction with state retention.

Power reduction without state retention culminates into a power switch, which switches off all combinatorial logic. To retain circuit states in e.g. flip-flops, while reducing weak-inversion currents, system states can either be stored in separate intrinsically low leakage, e.g. high threshold voltage, memory cells or the memory cell can be switched into a low leakage state. The latter can be done by:

- substrate biasing;

- source or gate biasing;

- Triple-S.

In Chapter 7 the effectiveness of weak-inversion current reduction techniques has been determined. It became clear that substrate biasing has no future in bulk CMOS processes, since the body factor diminishes, resulting in negligible leakage current reduction. As far as source and gate biasing is concerned, except for the application in power switches and the Triple-S mode switches, no practical solutions to implement both biasing techniques in sequential circuitry have been found yet. Triple-S, a new leakage current reduction technique, enables almost independent design of the leakage reduction factor and speed performance. Leakage reduction factors between 10^3 and 10^5 have been demonstrated for circuits containing flip-flops and latches. These experimental circuits have been processed in an adapted 0.25μm CMOS technology with low and high threshold voltages of 200mV and 600mV, respectively.

Chapter 8 presents the Triple-S technique applied to ring oscillators, latches, flip-flops, shift registers and binary counters, to determine speed, area, functional power and leakage. To implement the power reduction scheme of Triple-S in combination with power switches, the following is required:

- low- and high-threshold-voltage transistors;

- stable and defined clock signal during standby periods;

- "virtual" power supply lines and power switches.

Compared to "standard" designs, implementation of Triple-S and power switches requires on average 12% more chip area and consumes 4% more functional power. Propagation delays of Triple-S flip-flops increase by 10%. Overhead costs are dominant compared to the energy stored in the system. To regain the overhead costs, weak-inversion currents have to be reduced considerably and standby periods have to be large compared to active periods.

References

[1] www.semichips.org. Overall roadmap technology characteristics. Semiconductor Industry Association, 2000.

[2] Seymour Lipschutz. *Data Structures*. McGraw-Hill, 1986.

[3] A. Gersho and R. Gray. *Vector Quantization and Signal Compression*. Kluwer Academic Publishers, 1992.

[4] W.C. Fang, C. Y. Chang, and N.B.J. Sheu. A systolic tree-searched vector quantizer for real-time image compression. *Signal Processing IV*, 1992.

[5] C. Su, C. Tsui, and A. Despain. Low-power architecture design and compilation techniques for high-performance processors. *Compcon 1994*, pages 489–498, 1994.

[6] S. Raje et. al. Variable voltage sceduling. *International Symposium on Low-Power Design*, pages 9–14, 1995.

[7] K. Usami. Clustered voltage scaling technique for low-power design. *International Symposium on Low-Power Design*, pages 3–8, 1995.

[8] A. Chandrakasan and R. W. Brodersen. *Low-Power Digital CMOS Design*. Kluwer Academic Publishers, 1995.

[9] P. Macken. A voltage reduction technique for digital systems. *ISSCC 1990*, Februari 1990.

[10] V. Gutnik and A. Chandrakasan. An efficient controller for variable supply-voltage low-power processing. *VLSI Circuits Symposium*, pages 158–159, June 1996.

[11] N. Tzartzanis and W. Athas. Energy recovery for the design of high-speed, low-power static rams. *Proceedings of the international Symp.osium on Low-Power Electronics and Design*, pages 50–60, August 1996.

[12] T. Kuroda, T. Fujita, S. Mita, T. Nagamatsu, S. Yoshioka, K. Suzuki, F. Sano, M. Norishima, M. Murota, M. Kako, M. Kinugawa, M. Kakumu, and T. Sakurai. A 0.9-v 150-mhz, 10-mw, 4 mm2, 2-d discrete cosine. *IEEE Journal of Solid-State Circuits*, 31(11):1770–1777, November 1996.

[13] F. Assaderaghi. A dynamic threshold voltage mosfet (dtmos) for very low voltage operation. *IEEE Electron Device Letter*, 15(12):510–512, December 1994.

[14] L. Wei, Z. Chen, K. Roy, M.C. Johnson, Y. Ye, and V. K. De. Design and optimization of dual-threshold circuits for low-power applications. *IEEE Transactions on VLSI Systems*, 7(1):16–24, March 1999.

[15] R.P. Feynman. *Lectures on Computation*. Edison Wesley publishing compeny, 1996.

[16] C.H. Bennett. Logical reversibility of computation. *IBM journal of research and development*, 17:525–532, 1973.

[17] R. Landauer. Zig-zag path to understanding. In *Workshop on Physics and Computation*, pages 54–59. IEEE, 1994.

[18] R. Landauer. Irreversibility and heat generation in the computing process. *IBM J. Res. Dev.*, 5:183–191, 1961.

[19] H.S. Leff and A.F. Rex. *Maxwell's Demon: Entropy, Information, Computing*. Adam Hilger, 1990.

[20] T. D. Eastop and A. McConkey. *Applied Thermodynamics for Engineering Technologists*. Addison-Wesley, 1993.

[21] Yannis Tsividis. *Operation and Modeling of The MOS Transistor*. Number ISBN 0-07-065523-5. McGraw-Hill, second edition edition, 1999.

[22] J.A. Appels, E. Kooi, M.M. Paffen, J.J.H. Schatorje, and W.H.C.G. Verkuylen. Local oxidation of silicon and its application in semiconductor-device technology. *Philips Res. Rep.*, 25:118, 1970.

[23] A. Kamgar, S.J. Hillenius, R.M. Baker, S. Nakahara, and P.F. Bechtold. Gate oxide thinning at the active device/fox boundary in submicrometer pbl isolation. *IEEE Transactions on Electron Devices*, 42(12):2089–2095, December 1995.

[24] P. Sallagoity, M. Ada-Hanifi, M. Paoli, and M. Haond. Analysis of width edge effects in advanced isolation schemes for deep submicron cmos technologies. *IEEE Transactions on Electron Devices*, 43(11):1900 – 1906, November 1996.

[25] J.C.H. Hui, T.Y. Chiu, S. Weng, S. Wong, and W.G. Oldham. Sealed-interface local oxidation technology. *IEEE Transactions on Electron Devices*, ED-29(4):554–560, 1982.

[26] T. Speranza, Y. Wu, E. Fisch, J. Slinkman, J. Wong, and K. Beyer. Manufacturing optimization of shallow trench isolation for advanced cmos logic technology. *IEEE Advanced Semiconductor Manufacturing Conferece*, 2001.

[27] K. Ohe, S. Odanaka, K. Moriyama, T. Hori, and G. Fuse. Narrow-width effects of shallow trench-isolated cmos with n+ polysilicon gate. *IEEE Transactions on Electron Devices*, 36(6):1989, June 1989.

[28] A. Bryant, W. Haensch, S. Geissler, J. Mandelman, D. Poindexter, and M. Steger. The current-carrying corner inherent to trench isolation. *IEEE Electron Device Letters*, 14(8):412–414, August 1993.

[29] F.S. Shoucair. Scaling, subthreshold, and leakage current matching characteristics in high-temperature vlsi cmos devices. *IEEE Transcactions on Components, Hybrids, and Manufacturing Technology*, 12(4):780 – 788, December 1989.

[30] D.A. Neamen. *Semiconductor Physics and Devices*. Richard D. Irwin, Inc., ISBN 0-256-08405-X, Homewood, IL 60430, Boston, MA 02116, 1992.

[31] A. Keshavarzi, K.H. Roy, and F. Charles. Intrinsic leakage in low power deep submicron cmos ics. *Proceedings of the IEEE International Test Conference*, pages 146 – 155, 1997.

[32] R.F. Pierret. *Semiconductor Device Fundamentals*. Addison-Wesley, 1996.

[33] H. Brut and R.M.D.A. Velghe. Contribution to the characterization of the hump effect in mosfet submicrotronic technologies. In *Proceedings of the International Conference on Microelectronic Test Structures*, volume 12, pages 188 – 193. IEEE, March 1999.

[34] S.M. Sze. *High-Speed Semiconductor Devices*. John Wiley & Sons, Inc., New York, 1990.

[35] M. Rosar, B. Leroy, and G. Schweeger. A new model for the description of gate voltage and temperature dependence of gate induced drain leakage in the low electric field region. *IEEE Transactions on Electron Devices*, 47(1):154 – 159, Januari 2000.

[36] N. Yang, W.K. Henson, J.R. Hauser, and J.J. Wortman. Modeling study of ultrathin gate oxides using direct tunneling current and capacitance-voltage measurements in mos devices. *IEEE Transactions on Electron Devices*, 46(7):1464 – 1471, July 1999.

[37] N. Yang, W.K. Henson, and J. J. Wortman. A comparative study of gate direct tunneling and drain leakage currents in n-mosfet's with sub-2-nm gate oxides. *IEEE Transactions on Electron Devices*, 47(8):1636 – 1644, August 2000.

[38] H.J.M. Veendrick. Short-circuit dissipation in cmos circuitry and its impact on the design of buffer circuits. *IEEE Journal of Solid State Circuits*, 19(4):468 – 473, August 1984.

[39] E. Jacobs. Using gate sizing to reduce glitch power. In *IEEE Workshop on Circuits, Systems and Signal Processing*. Proceedings of the ProRISC, 1996.

[40] J. Sparso and J. Staunstrup. Delay-insensitive multi-ring structures. *Integration, the VLSI journal*, 15:313 – 340, 1993.

[41] L.S. Nielsen and C. Niessen. Low-power operation using self-timed circuits and adaptive scaling of the supply voltage. *IEEE Transactions on Very Large Scale Integration Systems*, 2(4), December 1994.

[42] Y. Nakagome, K. Itoh, M. Isoda, K. Takeuchi, and M. Aoki. Sub 1-v swing internal bus architecture for future low-power ulsi's. *IEEE Journal of Solid-State Circuits*, pages 414 – 419, April 1993.

[43] J. Kassakian, M. Schlecht, and G. Verghese. *Principles of Power Electronics*. Addison-Wesley, 1991.

[44] J. G. Koller and W.C. Athas. Adiabatic switching, low energy computing, and the physics of storing and erasing information. In *Proceedings of the Workshop on Physics and Computation*. IEEE, October 1992.

[45] L.J. Svensson and J.G. Koller. Adiabatic charging without inductors. In *Proceedings of the International Workshop on Low-Power Design*, pages 159 – 164. IEEE, April 1994.

[46] W.C. Athas, J.G. Koller, and L.J. Svensson. An energy-efficient cmos line driver using adiabatic switching. In *Proceedings of the Fourth Great Lakes Symposium on VLSI Design*, pages 159–164. IEEE, March 1994.

[47] Y. Ye and K. Roy. Reversible and quasi-static adiabatic logic. In *Proceedings of the ECCTD*. IEEE, September 1997.

[48] Y. Moon and D Jeong. A 32×32-b adiabatic register file with supply clock generator. *IEEE Journal of Solid-State Circuits*, May 1998.

[49] W. Athas, N. Tzartzanis, W. Mao, L. Peterson, R. Lal, K. Chong, J. Moon, L. Svensson, and M. Bolotski. The design and implementation of a low-power clock-powered microprocessor. *IEEE Journal of Solid-State Circuits*, 35(11):1561 – 1570, November 2000.

[50] Rusu S. Trends and challenges in vlsi technology scaling towards 100 nm , (invited). In *Proceedings of the ESSCIRC*, pages 23 –25. IEEE, September 2001.

[51] H. Yoshimura, Y. Asahi, and F. Matsuoka. Scaling senario of multi-level interconnects for future cmos lsi. In *Symposium on VLSI Technology Digest of Technical Papers*, pages 143 – 144. IEEE, 2001.

[52] K. Mistry, T. Ghani, M. Armstrong, S. Tyagi, P. Packan, S. Thompson, S. Yu, and M. Bohr. Scalability revisited: 100 nm pd-soi transistors and implications for 50 nm devices. In *Symposium on VLSI Technology Digest of Technical Papers*, pages 204 – 205. IEEE, 2000.

[53] J. D. Meindl and J. Davis. The fundamental limit on binary switching energy for terascale integration (tsi). *IEEE Journal of Solid-State Circuits*, 35(10):1515 – 1516, October 2000.

[54] K. Usami, M. Igarashi, F. Minami, T. Ishikawa, M. Kanazawa, M. Ichida, and K. Nogami. Automated low-power technique exploiting multiple supply voltages applied to a media processor. *IEEE Journal of Solid-State Circuits*, 33(3):463 – 472, March 1998.

[55] D. Ghosh and S. K. Nandy. Design and realization of high-performance wave-pipelined 8 $\times$ 8b multiplier in cmos technology. *IEEE Transactions on Very Large Scale Integration Systems*, 3(1):36 – 48, March 1995.

[56] V. von Kaenel, P. Macken, and M.G.R. Degrauwe. A voltage reduction technique for battery-operated systems. *IEEE Journal of Solid-State Circuits*, 25(5):1136 – 1140, October 1990.

[57] T. Kuroda, K. Suzuki, S. Mita, T. Fujita, F. Yamane, F. Sano, A. Chiba, Y. Watanabe, K. Matsuda, T. Maeda, T. Sakurai, and T. Furuyama. Variable supply-voltage scheme for low-power high-speed cmos digital design. *IEEE Journal of Solid-State Circuits*, 33(3):454 – 462, March 1998.

[58] P. T. Lai, J. P. Xu, W. M. Wong, H. B. Lo, and Y. C. Cheng. Correlation between hot-carrier-induced interface states and gidl current increase in n-mosfets. *IEEE Transactions on Electron Devices*, 45(2):521 – 528, Februari 1998.

[59] I. Chin, C.W. Teng, D.J. Coleman, and A. Nishimaru. Interface-trap enhanced gate-induced leakage current in mosfet. *IEEE Electron Device Letters*, 10(5):216 – 218, May 1989.

[60] S. Song, H.J. Kim, J.Y. Yoo, J.H. Yi, W.S. Kim, N.I. Lee, K. Fujihara, H.K. Kang, and J.T. Moon. On the gate oxide scaling of high performance cmos transistors. In *Proceedings of the IEDM 2001*, pages 55 – 58. IEEE, 2001.

[61] C.M. Osburn, I. Kim, S.K. Han, I. De, K.F. Yee, S. Gannavaram, S.J. Lee, C.H. Lee, Z.J. Luo, W. Zhu, J.R. Hauser, D.L. Kwong, G. Lucovsky, T.P. Ma, and M.C. Ozturk. Vertically scaled mosfet gate stacks and junctions: How far are we likely to go? *IBM journal of research and development*, 46(2/3):299 – 315, May 2002.

[62] S. Campbell, D.C. Gilmer, X. Wang, M. Hsieh, H. Kim, W.L Gladfelter, and J. Yan. Mosfet transistors fabricated with high permitivity tio_2 dielectrics. *IEEE Transactions on Electron Devices*, 44(1):104 – 109, Januari 1997.

[63] S.A. Campbell, H.S. Kim, D.C. Gilmer, B. He, T. Ma, and W.L. Gladfelter. Titanium dioxide based gate insulators. *IBM journal of research and development*, 43(3):383 – 392, May 1999.

[64] M. Houssa, M. Naili, V.V. Afanas'ev, M.M. Heyns, and A. Stesmans. Electrical and physical characterization of high-k dielectric layers. *IEEE*, 2001.

[65] C.H. Chen, Y.K. Fang, C.W. Yang, S.F. Ting, Y.S. Tsair, M.F. Wang, Y.M. Lin, M.C. Yu, S.C. Chen, C.H. Yu, and M.S. Liang. High-quality ultrathin (1.6 nm) nitride/oxide stack gate dielectric prepared by combining remote plasma nitridation and lpcvd technologies. *IEEE Electron Devide Letters*, 22(6):260 – 262, June 2001.

[66] J.A. Duffy. *Bonding Energy Levels and Bands in Inorganic Solids*. New York: Wiley, 1990.

[67] C.G. Parker, G. Lucovsky, and J.R. Hauser. Ultrathin oxide-nitride gate dielectric mosfet's. *IEEE Electron Device Letters*, 19(4):106 – 108, April 1998.

[68] C.H. Chen, Y.K. Fang, C.W. Yang, S.F. Ting, Y.S. Tsair, M.C. Yu, T.H. Hou, M.F. Wang, S.C. Chen, C.H. Yu, and M.S. Liang. Thermally-enhanced remote plasma nitrided ultrathin (1.65 nm) gate oxide with excellent performances in reduction of leakage current and boron diffusion. *IEEE Electron Device Letters*, 22(8):378 – 380, August 2001.

[69] P.R. van der Meer and A. van Staveren. New standby-current reduction technique for deep sub-micron vlsi cmos circuits: Smart series switch. In *Proceedings of the ESSCIRC*. IEEE, September 2002.

[70] S. Shigematsu, S. Mutoh, Y. Matsuya, Y. Tanabe, and J. Yamada. A 1-v high-speed mtcmos circuit scheme for power-down application circuits. *IEEE Journal of Solid-State Circuits*, 32(6):861 – 869, June 1997.

[71] N. Shibata, H. Morimura, and M. Harada. 1-v 100-mhz embedded sram techniques for battery-operated mtcmos/simox asics. *IEEE Journal of Solid-State Circuits*, 35(10):1396 – 1407, October 2000.

[72] Y. Oowaki, M. Noguchi, S. Takagi, D. Takashima, M. Ono, Y. Matsunaga, K. Sunouchi, H. Kawaguchiya, S. Matsuda, M. Kamoshida, T. Fuse, S. Watanabe, A. Toriumi, S. Manabe, and A. Hojo. A sub-0.1 μm circuit design with substrate-over-biasing. *IEEE International Solid State Circuit Conference*, pages 88 – 89, February 1998.

[73] N. Lindert, T. Sugii, S. Tang, and C. Hu. Dynamic threshold pass-transistor logic for improved delay at lower power supply voltages. *IEEE Journal of Solid-State Circuits*, 34(1):85 – 89, Januari 1999.

[74] P.R. van der Meer. Electronic digital circuit operable active mode and sleep mode. Patent application nr. 99203168.2, September 1999.

[75] P.R. van der Meer, A. van Staveren, and A.H.M. van Roermund. Ultra-low standby-currents for deep sub-micron vlsi cmos circuits: Smart series switch. In *Proceedings of the ISCAS*, volume 4, pages 1 – 4. IEEE, 2000.

[76] S. Mutoh, S. Shigematsu, Y. Gotoh, and S. Konaka. Design method of mtcmos power switch for low-voltage high-speed lsis. In *Proceedings of the Design Automation Conference*, volume 1, pages 113 – 116. IEEE, 1999.

[77] T. Douseki, S. Shigematsu, J. Yamada, M. Harada, H. Inokawa, and T. Tsuchiya. A 0.5-v mtcmos/simox logic gate. *IEEE Journal of Solid-State Circuits*, 32(10):1604 – 1609, October 1997.

[78] H. Kawaguchi, K. Nose, and T. Sakurai. A super cut-off cmos (sccmos) scheme for 0.5-v supply voltage with picoampere standby current. *IEEE Journal of Solid-State Circuits*, 35(10):1498 – 1501, October 2000.

[79] P.R. van der Meer and A. van Staveren. Effectivity of standby-enegy reduction techniques for deep sub-micron cmos. In *Proceedings of the ISCAS*. IEEE, May 2001.

[80] P.R. van der Meer. Sub-threshold current reduction: Trend in substrate biasing and the externally determined current technique. Technical report, Philips Semiconductors, December 1998.

[81] A.H. Montree, A.C.M.C. van Bradenburg, D.B.M. Klaassen, R.P. Llopis, Y.V. Ponomarev, R.F.M. Roes, A.J. Scholten, and R.S. van Veen. Limitations to adaptive back-bias approach for standby power reduction in deep submicron cmos. In *Proceedings of the ESSDERC*, page 580. IEEE, 1999.

[82] T. Kawahara, M. Horiguchi, Y. Kawajiri, G. Kitsukawa, T. Kure, and M. Aoki. Sub-threshold current reduction for decoded-driver by self-reverse biasing. *IEEE Journal of Solid-State Circuits*, 28(11):1136 – 1144, November 1993.

[83] M.C. Johnson, D. Somasekhar, L. Chiou, and K. Roy. Leakage control with efficient use of transistor stacks in single threshold cmos. *IEEE Transactions on VLSI Systems*, 10(1):1 – 5, Februari 2002.

Index